METHODS IN MOLECULAR BIOLOGY

Series Editor
John M. Walker
School of Life and Medical Sciences
University of Hertfordshire
Hatfield, Hertfordshire, UK

For further volumes:
http://www.springer.com/series/7651

For over 35 years, biological scientists have come to rely on the research protocols and methodologies in the critically acclaimed *Methods in Molecular Biology* series. The series was the first to introduce the step-by-step protocols approach that has become the standard in all biomedical protocol publishing. Each protocol is provided in readily-reproducible step-by-step fashion, opening with an introductory overview, a list of the materials and reagents needed to complete the experiment, and followed by a detailed procedure that is supported with a helpful notes section offering tips and tricks of the trade as well as troubleshooting advice. These hallmark features were introduced by series editor Dr. John Walker and constitute the key ingredient in each and every volume of the *Methods in Molecular Biology* series. Tested and trusted, comprehensive and reliable, all protocols from the series are indexed in PubMed.

Seed Dormancy

Methods and Protocols

Edited by

Naoto Kawakami

Department of Life Sciences, School of Agriculture, Meiji University, Kawasaki, Japan

Kazuhiro Sato

Institute of Plant Science & Resources, Okayama University, Kurashiki, Japan; Faculty of Agriculture, Setsunan University, Hirakata, Japan; Kazusa DNA Research Institute, Kisarazu, Japan

Humana Press

Editors
Naoto Kawakami
Department of Life Sciences, School of Agriculture
Meiji University
Kawasaki, Japan

Kazuhiro Sato
Institute of Plant Science & Resources
Okayama University
Kurashiki, Japan

Faculty of Agriculture
Setsunan University
Hirakata, Japan

Kazusa DNA Research Institute
Kisarazu, Japan

ISSN 1064-3745 ISSN 1940-6029 (electronic)
Methods in Molecular Biology
ISBN 978-1-0716-3964-1 ISBN 978-1-0716-3965-8 (eBook)
https://doi.org/10.1007/978-1-0716-3965-8

This Humana imprint is published by the registered company Springer Science+Business Media, LLC, part of Springer Nature.
The registered company address is: 1 New York Plaza, New York, NY 10004, U.S.A.

Preface

The previous edition of *Seed Dormancy*, published over 10 years ago, has been an informative and useful guide for studies of seed dormancy and germination. Recent advances in genome sequencing and editing technologies have facilitated the characterization and regulation of biological processes, especially in non-model wild and cultivated species. Seed dormancy is the result of adaptation of wild plants to conditions in nature; it has been inherited by cultivated crops in various forms during domestication, sometimes leading to severe problems in crop production.

This volume provides basic and state-of-the-art methods for fundamental seed research and practical application. The chapters in Part I introduce recent seed dormancy studies in model plants and crops, while the chapters in Part II present basic methods for germination and longevity assays applicable to various species, including parasitic plants. The seed is a living individual composed of the embryo, endosperm, and testa (seed coat), each of which has a considerable impact on dormancy, germination, and longevity. However, seed tissues are rich in storage materials and various secondary metabolites that often interfere with tissue imaging, biochemical, and molecular biological analyses. Part III presents tips for success in seed imaging, sugar, and chromatin analyses. Natural variation in seed dormancy and germination is a rich source of gene hunting for basic studies and plant breeding. Part IV describes details of QTL analysis and GWAS protocols for diploid and complex hexaploid species. Transient and stable transformation techniques for cereal species are also described. Considering that genetic transformation technologies are underdeveloped in non-model plants and crop species, we hope that these chapters will facilitate the development of new protocols for genetic transformation of various non-model plants. Finally, Part V discusses modern technologies and applications for pre-harvest sprouting-resistant wheat breeding in Japan, one of the most severe environments for wheat production. This volume complements the previous edition and will expand on basic and applied research, especially in non-model plants.

Kawasaki, Japan ***Naoto Kawakami***
Kurashiki, Japan ***Kazuhiro Sato***

Contents

Contributors

FUMITAKA ABE • *Institute of Crop Science, National Agriculture and Food Research Organization (NARO), Tsukuba, Ibaraki, Japan*
NANA ASHIKAGA • *Kitami Agricultural Experiment Station, Agricultural Research Department, Hokkaido Research Organization, Kunneppu, Hokkaido, Japan*
JULIA BUITINK • *Institut Agro, University of Angers, INRAE, IRHS, SFR QUASAV, Angers, France*
GILTSU CHOI • *Department of Biological Sciences, KAIST, Daejeon, Korea*
MAKIKO CHONO • *Institute of Crop Science, National Agriculture and Food Research Organization, Tsukuba, Ibaraki, Japan*
MASAYA FUJITA • *Institute of Crop Science, National Agriculture and Food Research Organization, Tsukuba, Ibaraki, Japan*
DENNYS-MARCELA GÓMEZ-PÁEZ • *Université Paris-Saclay, INRAE, AgroParisTech, Institut Jean-Pierre Bourgin (IJPB), Versailles, France*
EIKO HIMI • *School of Agriculture, Kibi International University, Minamiawaji, Japan*
HIROSHI HISANO • *Institute of Plant Science and Resources, Okayama University, Kurashiki, Japan*
ROBERT E. HOFFIE • *Leibniz Institute of Plant Genetics and Crop Plant Research (IPK) Gatersleben, Seeland, Germany*
RYOZO IMAI • *Division of Crop Genome Editing, Institute of Agrobiological Sciences, National Agriculture and Food Research Organization (NARO), Tsukuba, Japan*
YUJI ISHIDA • *Plant Innovation Center, Japan Tobacco Inc. (currently Agri-Bio Research Center, KANEKA CORPORATION), Iwata, Japan*
YOKO KAMIYA • *Kihara Institute for Biological Research, Yokohama City University, Yokohama, Japan*
NAOTO KAWAKAMI • *Department of Life Sciences, School of Agriculture, Meiji University, Kawasaki, Japan*
KANAKO KAWAURA • *Kihara Institute for Biological Research, Yokohama City University, Yokohama, Japan*
TOSHIHIKO KOMARI • *Plant Innovation Center, Japan Tobacco Inc. (currently Agri-Bio Research Center, KANEKA CORPORATION), Iwata, Japan*
JOCHEN KUMLEHN • *Leibniz Institute of Plant Genetics and Crop Plant Research (IPK) Gatersleben, Seeland, Germany*
SHIHO KURIHARA-YONEMOTO • *Hokkaido Agricultural Research Center, NARO, Sapporo, Japan*
YONGMIN KWON • *Department of Biological Sciences, KAIST, Daejeon, Korea*
OLIVIER LEPRINCE • *Institut Agro, University of Angers, INRAE, IRHS, SFR QUASAV, Angers, France*
TAO LI • *College of Life Sciences, Henan Agricultural University, Zhengzhou, China*
YUELIN LIU • *Institute of Plant and Food Science, Department of Biology, Southern University of Science and Technology, Shenzhen, China*
WEIFENG LUO • *Division of Crop Genome Editing, Institute of Agrobiological Sciences, National Agriculture and Food Research Organization (NARO), Tsukuba, Japan*

Enrico Magnani • *Université Paris-Saclay, INRAE, AgroParisTech, Institut Jean-Pierre Bourgin (IJPB), Versailles, France*
Hitoshi Matsunaka • *Hokkaido Agricultural Research Center, National Agriculture and Food Research Organization, Memuro, Hokkaido, Japan*
Masahiko Mori • *Institute of Crop Science, NARO, Tsukuba, Ibaraki, Japan; Obihiro University of Agriculture and Veterinary Medicine, Obihiro, Hokkaido, Japan*
Shingo Nakamura • *Institute of Crop Science, National Agriculture and Food Research Organization, Tsukuba, Ibaraki, Japan*
Masahiko Otani • *Department of Life Sciences, School of Agriculture, Meiji University, Kawasaki, Japan; Organization for the Strategic Coordination of Research and Intellectual Properties, Meiji University, Kawasaki, Japan*
Kazuhiro Sato • *Institute of Plant Science & Resources, Okayama University, Kurashiki, Japan; Faculty of Agriculture, Setsunan University, Hirakata, Japan; Kazusa DNA Research Institute, Kisarazu, Japan*
Yuichiro Tsuchiya • *Institute of Transformative Bio-Molecules, Nagoya University, Nagoya, Japan*
Jirui Wang • *Triticeae Research Institute, Sichuan Agricultural University, Chengdu, China*
Jian Yang • *Institute of Wheat, Henan Academy of Agricultural Sciences, Zhengzhou, China*
Jia Xin Yap • *Institute of Transformative Bio-Molecules, Nagoya University, Nagoya, Japan*
Tianyong Zhao • *State Key Laboratory for Crop Stress Resistance and High-Efficiency Production, College of Life Sciences, Northwest A&F University, Yangling, Shaanxi, China*
Lipeng Zheng • *Department of Life Sciences, School of Agriculture, Meiji University, Kawasaki, Japan; Institute of Health and Medicine, Hefei Comprehensive National Science Center, Hefei, China; School of Life Sciences, Division of Life Sciences and Medicine, University of Science and Technology of China, Anhui, China*

Part I

Introduction

Chapter 1

Genetic, Epigenetic, and Environmental Control of Seed Dormancy and Germination

Masahiko Otani, Lipeng Zheng, and Naoto Kawakami

Abstract

Seed germination is controlled by a combination of the seed dormancy level and environmental conditions such as light, temperature, moisture, and nitrate levels. Seed dormancy is programed genetically, but it is also sensitive to maternal environmental conditions before and after anthesis. Recent developments in molecular genetics and bioinformatics have greatly enhanced our understanding of the molecular mechanisms of seed dormancy and germination in model plants and economically important crop species. This chapter focuses on temperature as an environmental factor and discusses the genetic and epigenetic mechanisms of dormancy and germination.

Key words Chromatin remodeling, Phytohormone, Temperature, Ttranscription network

Seed dormancy is defined as the failure of an intact viable seed to complete germination under favorable conditions. Types of dormancy vary among species and are categorized as physiological, morphological, or physical dormancy and their combinations [1]. Physiological dormancy is caused by physiological inhibition of the seed that prevents radicle emergence and is common in the seeds of most weeds and vegetables [1]. This chapter reviews the molecular mechanisms of the genetic and environmental regulation of physiological dormancy and germination, focusing on the effects of temperature.

1 Regulation of Phytohormone Metabolism and Signaling by Light and Temperature

Abscisic acid (ABA) and gibberellins (GAs) are the main phytohormones involved in germination control. Light-induced germination of photoblastic seeds has long been studied, and the first exciting discovery was the plant photoreceptor, the phytochrome [2]. Phytochrome B (PhyB) controls the red/far-red reversible

Naoto Kawakami and Kazuhiro Sato (eds.), *Seed Dormancy: Methods and Protocols*, Methods in Molecular Biology, vol. 2830,
https://doi.org/10.1007/978-1-0716-3965-8_1,

germination response, while PhyA photoirreversibly triggers germination upon irradiation with ultraviolet, visible, and far-red light [3]. In lettuce (*Lactuca sativa*) and Arabidopsis (*Arabidopsis thaliana*) seeds, GA levels are regulated by phytochrome at the transcriptional level of GA metabolism genes. In both species, genes encoding the GA biosynthesis enzyme GA 3-oxidase (*GA3ox*) and GA catabolizing enzyme GA 2-oxidase (*GA2ox*) are regulated in a photoreversible manner [4, 5]. ABA levels are also regulated by phytochrome at the transcriptional level of ABA biosynthesis and catabolism enzyme genes in Arabidopsis seeds [6]. A set of basic Helix-Loop-Helix (bHLH) transcription factors called PHYTOCHROME-INTERACTING FACTORs (PIFs) negatively regulates light signaling. The light-absorbed active form of phytochrome, Pfr, is transported to the nucleus where it specifically binds to PIF proteins, and then the PIFs are degraded via the ubiquitin–proteasome pathway. In Arabidopsis seeds, PIF1 (also known as PIL5) negatively regulates seed germination by activating the transcription of *SOMNUS* (*SOM*), which indirectly regulates ABA and GA metabolism enzyme genes [7, 8]. The Pfr-induced degradation of PIF1/PIL5 deactivates *SOM* expression, and germination is induced by the suppression of ABA biosynthesis and induction of GA biosynthesis.

The germination temperature range is closely related to the level of dormancy in winter and summer annual plants [1]. Freshly dispersed and soil-buried seeds of obligate winter annuals do not germinate at any temperature, even under illumination in spring, and the seeds change sensitivity to temperature during after-ripening from spring to autumn. The maximum permissive temperature for germination rises gradually, and the seeds germinate in autumn when the temperature falls below the upper limit [1]. By contrast, the germination of summer annual seeds is inhibited by low temperatures until early spring. The minimum permissive temperature for germination drops from autumn to spring, and seeds germinate when the soil temperature rises above this limit in spring. A change in seed sensitivity to temperature during after-ripening is also observed in Arabidopsis seeds harvested and stored under laboratory conditions [9].

In imbibed lettuce and Arabidopsis seeds, the ABA content is under the control of temperature, and de novo ABA biosynthesis is required for germination inhibition at supraoptimal high temperatures (thermoinhibition) [10, 11]. In both species, thermoinhibition is caused by upregulation of ABA biosynthesis genes (*NCED*) and suppression of ABA catabolism (*CYP707A*) and GA biosynthesis (*GA3ox*) genes at supraoptimal temperatures [11, 12]. Therefore, the regulation of hormone metabolism gene expression plays a critical role in the germination response to temperature, as observed in light-induced seed germination.

The photoreceptor phytochrome has recently been reported to function as a temperature sensor in Arabidopsis seedlings. High temperatures induce thermomorphogenesis, including elongation of the hypocotyl and petiole and upward growth of rosette leaves. These morphological changes enhance the air flow between leaves and allow the leaf temperature to cool [13]. In this process, the phytochrome-interacting factor PIF4 plays a central role in auxin production and cell wall modification. An increased rate of reversion of phytochrome from active Pfr to inactive Pr at high temperatures (thermal reversion) allows the accumulation of PIF4 and thermomorphogenesis [14, 15]. Notably, *SOM* expression is induced at supraoptimal temperatures in red-light-irradiated seeds imbibed in the dark, and SOM is responsible for the thermoinhibition [16]. Phytochromes may also work as thermosensors in seeds and are involved in the regulation of *SOM* in response to temperature. Recently, it was shown that PhyB in the Arabidopsis seed endosperm acts as a thermosensor, and ABA metabolism is regulated by PIF3 downstream from PhyB [17].

2 Novel Mechanisms Regulating ABA Signaling in Seeds

Quantitative trait loci (QTL) that control seed dormancy have been analyzed in cereal species and Arabidopsis, and causal genes related to signal transduction, transcriptional regulation, and amino acid metabolism have been identified (*see* Chap. 2) [18–24]. *Delay of Germination1* (*DOG1*) is the master regulator of seed dormancy and was identified from a dormancy QTL analysis of Arabidopsis [18, 25]. Recently, it was revealed that DOG1 enhances ABA signaling independently of ABA by binding and inactivating protein phosphatase 2Cs, which are negative regulators of ABA signaling [23, 24].

Another novel regulator of ABA signaling was discovered by analyzing the seed germination response to temperature in rice [26]. A genome-wide association study (GWAS) of 164 cultivars of japonica rice with genome sequence and germination data at low and optimal temperatures revealed the 14-3-3 protein coding gene *GF14h*. These proteins can bind to the phosphorylated motifs of target proteins and modulate their function. GF14h suppresses ABA signaling and promotes germination at optimal temperatures. Interestingly, 10–20% of Japanese landraces and cultivars developed before 1990 retain a functional *GF14h* allele, but most of the modern cultivars in northern Japan and China have nonfunctional GF14h alleles. The loss of GF14h function in modern cultivars may be the result of breeding for preharvest sprouting resistance [26].

3 Phase Transition from Seed Maturation to Germination and Vegetative Growth

Arabidopsis mutant analyses have shown that *LAFL* genes (*LEC1*, *LEAFY COTYLEDON1*; *ABI3*, *ABSCISIC ACID INSENSITIVE3*; *FUS3*, *FUSCA3*; and *LEC2*) act as master regulators of seed maturation programs, including dormancy development [27, 28]. The seeds of loss-of-function mutants of *LAFL* undergo viviparous germination. *LEC1* is a homolog of the mammalian transcription factor NF-YB [29–31], while *ABI3*, *FUS3*, and *LEC2* are plant-specific B3 domain transcription factors [32–35]. These LAFL proteins have partially overlapping synergistic functions, and they can form multiprotein complexes to activate their target genes [36, 37]. LAFL enhances *DOG1* expression by inducing *Leucine Zipper Transcription Factor67* (*bZIP67*) expression, which transactivates *DOG1* [38].

Seed dormancy 4 (*Sdr4*) is a positive regulator of dormancy development based on QTL analysis of rice preharvest sprouting resistance [19]. Sdr4 is a nuclear localized master regulator of seed dormancy in rice. Arabidopsis has seven *Sdr4* homologs in its genome, and the most homologous gene is designated *Sdr Four Like1* (*SFL1*) [19]. Unlike rice *Sdr4*, *SFL1* (also known as *AtSdr4L* and *ODR1*) positively regulates germination and postgerminative growth in Arabidopsis [39–41]. Recently, *SFL1* was shown to regulate the timing of *LAFL* expression, and positively regulate dormancy development at the mid-maturation stage in the same way as rice *Sdr4* [41]. Therefore, *SFL1* is a unique regulator of the phase transition from seed maturation to germination and vegetative growth.

4 Epigenetic Regulation of Seed Dormancy and Germination

Molecular genetics and pharmacological studies indicate that epigenetic regulation is also critical for seed dormancy and germination. Arabidopsis reduced-dormancy mutant analyses revealed that *Reduced Dormancy2* (*RDO2*) encodes transcription elongation factor SII (TFIIS) [42, 43], and *RDO4/HUB1* encodes the C3HC4 RING finger proteins necessary for histone H2B monoubiquitination [44]. The RDO2 and RDO4/HUB1 proteins have been predicted to interact with the RNA polymerase II-associated factor 1 complex (PAF1C) and promote transcription elongation of several dormancy-related genes [43].

Trichostatin A (TSA), an inhibitor of histone deacetylase (HDAC), inhibits seed germination and seedling development [45]. Suppression of *LAFL* function during seed germination is necessary for postgerminative growth [46]. TSA treatment induces ectopic expression of *LAFL* in growth-arrested seedlings, and

HDAC coding *HDA6* gene knock-down mutant seedlings also show ectopic *LAFL* expression and embryo-like structures in the seedlings. GWAS of 113 natural accessions of Arabidopsis identified the HDAC gene *HD2B* as a modulator of seed dormancy and germination [47].

In addition to enhanced seed dormancy, *SFL1* loss-of-function mutant seedlings have a growth-arrest phenotype [41], implying that *SFL1* is a novel suppressor of embryonic characteristics, such as the chromatin-remodeling-related factors *PKL*, *HSI/VAL*, *HDA6*, and *HDA19* [48–50]. *LAFL* repression is mediated by HSI/VAL via recruitment of chromatin remodeling factors [46, 49, 50], and a protein–protein interaction between one of the *SFL1* paralogs (SFL2) and HSI/VAL has been suggested from yeast two-hybrid analysis [51]. HSI/VAL associates with HDA6, HDA19, and the Polycomb Repressive Complexes PRC1 and PRC2 [27, 45]. SFL1 and its paralogs may be involved in the stable repression of *LAFL* and promote germination and postgerminative growth [41].

5 Maternal Temperature and Seed Germination

The temperatures during seed maturation often correlate with the dormancy levels of mature seeds. In Arabidopsis, cool conditions during seed maturation enhance *DOG1* expression by increasing bZIP67 protein levels, resulting in enhanced seed dormancy [38]. *MOTHER OF FT AND TFL1* (*MFT*) is responsible for the low-temperature enhancement of dormancy during wheat grain development [52]. Maternal temperatures experienced during the vegetative stage also influence seed dormancy and the germination response to temperature. Mature seeds from wild oat (*Avena fatua*) plants that experienced low temperatures before anthesis showed increased dormancy, especially when the seeds were imbibed at low temperature for the dormancy test [53]. In Arabidopsis rosette leaves, *FLOWERING LOCUS C* (*FLC*) negatively regulates flowering genes such as *FT*, and confers a vernalization requirement. Long-term cold treatment (vernalization) induces expression of antisense transcripts at the *FLC* loci *COOLAIR*, which facilitate the silencing of *FLC* by recruiting PRC2 [54]. In Arabidopsis, low temperatures before anthesis increase seed dormancy by decreasing *FT* expression in the silique phloem [55]. Contrasting vernalization, *FLC* acts downstream from *FT* and promotes seed dormancy development in response to cold. FT suppresses *FLC* expression by activating *COOLAIR* transcription [56]. Low temperatures experienced at the rosette stage before bolting also influence the dormancy of Arabidopsis seeds [57].

6 Molecular Dissection of Dormancy Cycling and Germination Timing

Nondormant soil-buried seeds may acquire secondary dormancy in unfavorable conditions for germination and show an annual cycle of dormancy status between deep and shallow dormancy [1, 58]. This dormancy cycle allows winter and summer annual seeds in the soil seed bank to germinate at the appropriate time of year. Seeds in the deep dormancy stage are insensitive to external and internal germination-stimulating signals, such as light, nitrate, and GA, but are sensitive to ABA. By contrast, seeds in the shallow dormancy stage are sensitive to germination-stimulating signals and are relatively insensitive to ABA [1]. A molecular dissection of dormancy cycling has been reported that analyzed gene expression in Arabidopsis seeds [59, 60]. Soil-buried seeds of the winter annual ecotype Cvi showed distinct expression of hormone, nitrate, and light-sensitivity genes between the deep and shallow dormancy seasons. In deep dormant seeds, seed maturation and dormancy regulator genes (e.g., *DOG1* and *ABI3*) are expressed, and *DOG1* has a central role in the depth of dormancy by enhancing ABA signaling [58, 59]. In comparison, in shallow dormant seeds, expression of DELLA (RGL2), a negative regulator of GA signaling, may suppress germination under unfavorable conditions, while germination signals such as light can remove the DELLA suppression by stimulating GA biosynthesis.

QTLs for germination timing during the year (seedling emergence timing, SET) were analyzed using a mapping population from crosses between winter (Cvi) and summer (Bur) annual ecotypes of Arabidopsis [61]. Four out of five *SET* QTLs collocated with dormancy QTLs (*DOG20*, *DOG21*, *DOG22*, and *DOG6*). The most significant QTL, *SET1*, did not collocate with a *DOG*. Candidate genes in each locus were narrowed down by analyzing the annual gene expression profile and gene ontology, and two proteins that work upstream and downstream of DOG1 (*PDF1* and *AHG1*) and a membrane-associated NAC transcription factor (*ANAC060*) were identified [61]. AHG1 is a PP2C family protein that negatively regulates ABA signaling, while DOG1 enhances ABA signaling by inactivating AHG1 [23]. PDF1 is a subunit of protein phosphatase 2A and plays a negative role in ABA sensitivity and seed dormancy upstream from DOG1 [23]. *ANAC060* has been identified as a likely candidate gene for *DOG6*, which collocates with *SET4* [62]. ANAC060 is a negative regulator of ABA sensitivity and seed dormancy [63]. These observations suggest that germination timing is not under the control of DOG1, but that the negative regulation of ABA signaling is critical for dormancy relief and germination timing [58, 61].

7 Perspectives

Seed dormancy is an adaptive trait that enables plants to survive adverse conditions and restart growth in a season and location suitable for vegetative and reproductive growth. Manipulation of dormancy is important for crop production, since dormancy interferes with the rapid uniform germination of seeds and weak dormancy induces premature germination, which reduces the quality and yield of the seeds. The rapid development of molecular genetics and bioinformatics technologies has greatly enhanced our understanding of the molecular mechanisms of seed dormancy and germination using model plants and economically important crop species. These methods can be applied to diverse plant species. Natural and induced genetic variation in wild and cultivated species is a rich source of information on seed dormancy and germination, and some fruits from a few plant species are described in this chapter. The combined use of these technologies with genome editing and artificial intelligence will accelerate basic studies and the application of knowledge to crop production and coping with climate change.

References

1. Baskin JM, Baskin CC (2014) Seeds. Ecology, biogeography, and evolution of dormancy and germination, 2nd edn. Academic Press, San Diego
2. Borthwick HA, Hendricks SB, Parker MW, Toole EH, Toole VK (1952) A reversible photoreaction controlling seed germination. Proc Natl Acad Sci USA 38:662–666
3. Shinomura T, Nagatani A, Hanzawa H, Kubota M, Watanabe M, Furuya M (1996) Action spectra for phytochrome A- and B-specific photoinduction of seed germination in *Arabidopsis thaliana*. Proc Natl Acad Sci USA 93:8129–8133
4. Toyomasu T, Kawaide H, Mitsuhashi W, Inoue Y, Kamiya Y (1998) Phytochrome regulates gibberellin biosynthesis during germination of photoblastic lettuce seeds. Plant Physiol 118:1517–1523
5. Yamaguchi S, Smith MW, Brown RGS, Kamiya Y, Sun TP (1998) Phytochrome regulation and differential expression of gibberellin 3β-hydroxylase genes in germinating Arabidopsis seeds. Plant Cell 10:2115–2126
6. Seo M, Hanada A, Kuwahara A, Endo A, Okamoto M, Yamauchi Y, North H, Marion-Poll A, Sun TP, Koshiba T, Kamiya Y, Yamaguchi S, Nambara E (2006) Regulation of hormone metabolism in Arabidopsis seeds: phytochrome regulation of abscisic acid metabolism and abscisic acid regulation of gibberellin metabolism. Plant J 48:354–366
7. Oh E, Kim J, Park E, Kim JI, Kang C, Choi G (2004) PIL5, a phytochrome-interacting basic helix-loop-helix protein, is a key negative regulator of seed germination in *Arabidopsis thaliana*. Plant Cell 16:3045–3058
8. Kim DH, Yamaguchi S, Lim S, Oh E, Park J, Hanada A, Kamiya Y, Choi G (2008) SOMNUS, a CCCH-type zinc finger protein in *Arabidopsis*, negatively regulates light-dependent seed germination downstream of PIL5. Plant Cell 20:1260–1277
9. Tamura N, Yoshida T, Tanaka A, Sasaki R, Bando A, Toh S, Lepiniec L, Kawakami N (2006) Isolation and characterization of high temperature-resistant germination mutants of *Arabidopsis thaliana*. Plant Cell Physiol 47: 1081–1094
10. Yoshioka T, Endo T, Satoh S (1998) Restoration of seed germination at supraoptimal temperatures by fluridone, an inhibitor of abscisic acid biosynthesis. Plant Cell Physiol 39:307–312
11. Toh S, Imamura A, Watanabe A, Nakabayashi K, Okamoto M, Jikumaru Y, Hanada A, Aso Y, Ishiyama K, Tamura N, Iuchi S, Kobayashi M, Yamaguchi S,

Kamiya Y, Nambara E, Kawakami N (2008) High temperature-induced abscisic acid biosynthesis and its role in the inhibition of gibberellin action in Arabidopsis seeds. Plant Physiol 146:1368–1385

12. Argyris J, Dahal P, Hayashi E, Still DW, Bradford KJ (2008) Genetic variation for lettuce seed thermoinhibition is associated with temperature-sensitive expression of abscisic acid, gibberellin, and ethylene biosynthesis, metabolism, and response genes. Plant Physiol 148:926–947
13. Quint M, Delker C, Franklin KA, Wigge PA, Halliday KJ, van Zanten M (2016) Molecular and genetic control of plant thermomorphogenesis. Nat Plants 2:15190
14. Jung JH, Domijan M, Klose C, Biswas S, Ezer D, Gao M, Khattak AK, Box MS, Charoensawan V, Cortijo S, Kumar M, Grant A, Locke JCW, Schäfer E, Jaeger KE, Wigge PA (2016) Phytochromes function as thermosensors in *Arabidopsis*. Science 354: 886–889
15. Legris M, Klose C, Burgie ES, Rojas CCR, Neme M, Hiltbrunner A, Wigge PA, Schäfer E, Vierstra RD, Casal JJ (2016) Phytochrome B integrates light and temperature signals in *Arabidopsis*. Science 354:897–900
16. Lim S, Park J, Lee N, Jeong J, Toh S, Watanabe A, Kim J, Kang H, Kim DH, Kawakami N, Choi G (2013) ABA-INSENSITIVE3, ABA-INSENSITIVE5, and DELLAs interact to activate the expression of *SOMNUS* and other high-temperature-inducible genes in imbibed seeds in *Arabidopsis*. Plant Cell 25:4863–4878
17. Piskurewicz U, Sentandreu M, Iwasaki M, Glauser G, Lopez-Molina L (2023) The Arabidopsis endosperm is a temperature-sensing tissue that implements seed thermoinhibition through PhyB. Nat Commun 14:1202
18. Bentsink L, Jowett J, Hanhart CJ, Koornneef M (2006) Cloning of *DOG1*, a quantitative trait locus controlling seed dormancy in *Arabidopsis*. Proc Natl Acad Sci USA 103:17042–17047
19. Sugimoto K, Takeuchi Y, Ebana K, Miyao A, Hirochika H, Hara N, Ishiyama K, Kobayashi M, Ban Y, Hattori T, Yano M (2010) Molecular cloning of *Sdr4*, a regulator involved in seed dormancy and domestication of rice. Proc Natl Acad Sci USA 107:5792–5797
20. Torada A, Koike M, Ogawa T, Takenouchi Y, Tadamura K, Wu J, Matsumoto T, Kawaura K, Ogihara Y (2016) A causal gene for seed dormancy on wheat chromosome 4A encodes a MAP kinase kinase. Curr Biol 26:782–787
21. Nakamura S, Pourkheirandish M, Morishige H, Kubo Y, Nakamura M, Ichimura K, Seo S, Kanamori H, Wu J, Ando T, Hensel G, Sameri M, Stein N, Sato K, Matsumoto T, Yano M, Komatsuda T (2016) *Mitogen-activated protein kinase kinase 3* regulates seed dormancy in barley. Curr Biol 26:775–781
22. Sato K, Yamane M, Yamaji N, Kanamori H, Tagiri A, Schwerdt JG, Fincher GB, Matsumoto T, Takeda K, Komatsuda T (2016) Alanine aminotransferase controls seed dormancy in barley. Nat Commun 7:11625
23. Nee G, Kramer K, Nakabayashi K, Yuan B, Xiang Y, Miatton E, Finkemeier I, Soppe WJJ (2017) DELAY OF GERMINATION1 requires PP2C phosphatases of the ABA signalling pathway to control seed dormancy. Nat Commun 8:72
24. Nishimura N, Tsuchiya W, Moresco JJ, Hayashi Y, Satoh K, Kaiwa N, Irisa T, Kinoshita T, Schroeder JI, Yates JR, Hirayama T, Yamazaki T (2018) Control of seed dormancy and germination by DOG1-AHG1 PP2C phosphatase complex via binding to heme. Nat Commun 9:2132
25. Alonso-Blanco C, Bentsink L, Hanhart CJ, Blankestijn-de Vries H, Koornneef M (2003) Analysis of natural allelic variation at seed dormancy loci of *Arabidopsis thaliana*. Genetics 164:711–729
26. Yoshida H, Hirano K, Yano K, Wang F, Mori M, Kawamura M, Koketsu E, Hattori M, Ordonio RL, Huang P, Yamamoto E, Matsuoka M (2022) Genome-wide association study identifies a gene responsible for temperature-dependent rice germination. Nat Commun 13:5665
27. Lepiniec L, Devic M, Roscoe TJ, Bouyer D, Zhou DX, Boulard C, Baud S, Dubreucq B (2018) Molecular and epigenetic regulations and functions of the LAFL transcriptional regulators that control seed development. Plant Reprod 31:291–307
28. Jo L, Pelletier JM, Harada JJ (2019) Central role of the LEAFY COTYLEDON1 transcription factor in seed development. J Integr Plant Biol 61:564–580
29. Meinke DW (1992) A homoeotic mutant of *Arabidopsis thaliana* with Leafy Cotyledons. Science 258:1647–1650
30. Lotan T, Ohto M, Yee KM, West MA, Lo R, Kwong RW, Yamagishi K, Fischer RL, Goldberg RB, Harada JJ (1998) *Arabidopsis* LEAFY COTYLEDON1 is sufficient to induce embryo development in vegetative cells. Cell 93:1195–1205

31. Harada JJ (2001) Role of *Arabidopsis LEAFY COTYLEDON* genes in seed development. J Plant Physiol 158:405–409
32. Giraudat J, Hauge BM, Valon C, Smalle J, Parcy F, Goodman HM (1992) Isolation of the Arabidopsis ABI3 gene by positional cloning. Plant Cell 4:1251–1261
33. Luerssen H, Kirik V, Herrmann P, Misera S (1998) *FUSCA3* encodes a protein with a conserved VP1/AB13-like B3 domain, which is of functional importance for the regulation of seed maturation in *Arabidopsis thaliana*. Plant J 15:755–764
34. Stone SL, Kwong LW, Yee KM, Pelletier J, Lepiniec L, Fischer RL, Goldberg RB, Harada JJ (2001) *LEAFY COTYLEDON2* encodes a B3 domain transcription factor that induces embryo development. Proc Natl Acad Sci USA 98:11806–11811
35. Pelletier JM, Kwong RW, Park S, Le BH, Baden R, Cagliari A, Hashimoto M, Munoz MD, Fischer RL, Goldberg RB, Harada JJ (2017) LEC1 sequentially regulates the transcription of genes involved in diverse developmental processes during seed development. Proc Natl Acad Sci USA 114:6710–6719
36. Baud S, Kelemen Z, Thévenin J, Boulard C, Blanchet S, To A, Payre M, Berger N, Effroy-Cuzzi D, Franco-Zorrilla JM, Godoy M, Solano R, Thevenon E, Parcy F, Lepiniec L, Dubreucq B (2016) Deciphering the molecular mechanisms underpinning the transcriptional control of gene expression by master transcriptional regulators in Arabidopsis seed. Plant Physiol 171:1099–1112
37. Boulard C, Thévenin J, Tranquet O, Laporte V, Lepiniec L, Dubreucq B (2018) LEC1 (NF-YB9) directly interacts with LEC2 to control gene expression in seed. Biochim Biophys Acta Gene Regul Mech 1861:443–450
38. Bryant FM, Hughes D, Hassani-Pak K, Eastmond PJ (2019) Basic LEUCINE ZIPPER TRANSCRIPTION FACTOR67 transactivates *DELAY OF GERMINATION1* to establish primary seed dormancy in Arabidopsis. Plant Cell 31:1276–1288
39. Cao H, Han Y, Li J, Ding M, Li Y, Li X, Chen F, Soppe WJ, Liu Y (2020) *Arabidopsis thaliana* SEED DORMANCY 4-LIKE regulates dormancy and germination by mediating the gibberellin pathway. J Exp Bot 71:919–933
40. Liu F, Zhang H, Ding L, Soppe WJJ, Xiang Y (2020) REVERSAL OF RDO5 1, a homolog of Rice Seed Dormancy4, interacts with bHLH57 and controls ABA biosynthesis and seed dormancy in Arabidopsis. Plant Cell 32: 1933–1948
41. Zheng L, Otani M, Kanno Y, Seo M, Yoshitake Y, Yoshimoto K, Sugimoto K, Kawakami N (2022) Seed dormancy 4 like1 of Arabidopsis is a key regulator of phase transition from embryo to vegetative development. Plant J 112:460–475
42. Grasser M, Kane CM, Merkle T, Melzer M, Emmersen J, Grasser KD (2009) Transcript elongation factor TFIIS is involved in *Arabidopsis* seed dormancy. J Mol Biol 386:598–611
43. Liu Y, Geyer R, van Zanten M, Carles A, Li Y, Horold A, van Nocker S, Soppe WJJ (2011) Identification of the Arabidopsis *REDUCED DORMANCY 2* gene uncovers a role for the polymerase associated factor 1 complex in seed dormancy. PLoS One 6:e22241
44. Liu Y, Koornneef M, Soppe WJJ (2007) The absence of histone H2B monoubiquitination in the *Arabidopsis hub1* (*rdo4*) mutant reveals a role for chromatin remodeling in seed dormancy. Plant Cell 19:433–444
45. Tanaka M, Kikuchi A, Kamada H (2008) The Arabidopsis histone deacetylases HDA6 and HDA19 contribute to the repression of embryonic properties after germination. Plant Physiol 146:149–161
46. Jia H, Suzuki M, McCarty DR (2014) Regulation of the seed to seedling developmental phase transition by the LAFL and VAL transcription factor networks. Wiley Interdiscip Rev Dev Biol 3:135–145
47. Yano R, Takebayashi Y, Nambara E, Kamiya Y, Seo M (2013) Combining association mapping and transcriptomics identify *HD2B* histone deacetylase as a genetic factor associated with seed dormancy in *Arabidopsis thaliana*. Plant J 74:815–828
48. Ogas J, Kaufmann S, Henderson J, Somerville C (1999) PICKLE is a CHD3 chromatin-remodeling factor that regulates the transition from embryonic to vegetative development in *Arabidopsis*. Proc Natl Acad Sci USA 96:13839–13844
49. Tsukagoshi T, Morikami A, Nakamura K (2007) Two B3 domain transcriptional repressors prevent sugar-inducible expression of seed maturation genes in *Arabidopsis* seedlings. Proc Natl Acad Sci USA 104:2543–2547
50. Suzuki M, Wang HH, McCarty DR (2007) Repression of the *LEAFY COTYLEDON 1/B3* regulatory network in plant embryo development by *VP1/ABSCISIC ACID INSENSITIVE 3-LIKE* B3 genes. Plant Physiol 143:902–911
51. Altmann M, Altmann S, Rodriguez PA, Weller B, Elorduy Vergara L, Palme J, Marín-de la Rosa N, Sauer M, Wenig M, Villaécija-Aguilar

JA, Sales J, Lin CW, Pandiarajan R, Young V, Strobel A, Gross L, Carbonnel S, Kugler KG, Garcia-Molina A, Bassel GW, Falter C, Klaus Mayer KFX, Gutjahr C, Vlot AC, Grill E, Falter-Braun P (2020) Extensive signal integration by the phytohormone protein network. Nature 583:271–276

52. Nakamura S, Abe F, Kawahigashi H, Nakazono K, Tagiri A, Matsumoto T, Utsugi S, Ogawa T, Handa H, Ishida H, Mori M, Kawaura K, Ogihara Y, Miura H (2011) A wheat homolog of MOTHER OF FT AND TFL1 acts in the regulation of germination. Plant Cell 23:3215–3229
53. Sawhney R, Quick WA, Hsiao AI (1985) The effect of temperature during parental vegetative growth on seed-germination of wild oats (*Avena fatua* L). Ann Bot 55:25–28
54. Csorba T, Questa JI, Sun QW, Dean C (2014) Antisense COOLAIR mediates the coordinated switching of chromatin states at FLC during vernalization. Proc Natl Acad Sci USA 111:16160–16165
55. Chen M, MacGregor DR, Dave A, Florance H, Moore K, Paszkiewicz K, Smirnoff N, Graham IA, Penfield S (2014) Maternal temperature history activates Flowering Locus T in fruits to control progeny dormancy according to time of year. Proc Natl Acad Sci USA 111: 18787–18792
56. Chen M, Penfield S (2018) Feedback regulation of COOLAIR expression controls seed dormancy and flowering time. Science 360: 1014–1017
57. Auge GA, Blair LK, Neville H, Donohue K (2017) Maternal vernalization and vernalization-pathway genes influence progeny seed germination. New Phytol 216:388–400
58. Finch-Savage WE, Footitt S (2017) Seed dormancy cycling and the regulation of dormancy mechanisms to time germination in variable field environments. J Exp Bot 68:843–856
59. Footitt S, Douterelo-Soler I, Clay H, Finch-Savage WE (2011) Dormancy cycling in *Arabidopsis* seeds is controlled by seasonally distinct hormone-signaling pathways. Proc Natl Acad Sci USA 108:20236–20241
60. Footitt S, Huang ZY, Clay HA, Mead A, Finch-Savage WE (2013) Temperature, light and nitrate sensing coordinate Arabidopsis seed dormancy cycling, resulting in winter and summer annual phenotypes. Plant J 74:1003–1015
61. Footitt S, Walley PG, Lynn JR, Hambidge AJ, Penfield S, Finch-Savage WE (2020) Trait analysis reveals DOG1 determines initial depth of seed dormancy, but not changes during dormancy cycling that result in seedling emergence timing. New Phytol 225:2035–2047
62. Song S, He H, Gühl K, van Bolderen-Veldkamp M, Buijs G, Willems LAJ, Bentsink L (2021) DELAY OF GERMINATION 6, encoding the ANAC060 transcription factor, inhibits seed dormancy. bioRxiv. https://doi.org/10.1101/2021.05.03.442418
63. Song S, Willems JAJ, Jiao A, Zhao T, Schrenz ME, Bentsink L (2022) The membrane associated NAC transcription factors ANAC060 and ANAC040 are functionally redundant in the inhibition of seed dormancy in *Arabidopsis thaliana*. J Exp Bot 73:5514–5528

Chapter 2

Regulation of Seed Dormancy Genes in Triticeae Species

Kazuhiro Sato, Shingo Nakamura, and Masaya Fujita

Abstract

Wild progenitors of Triticeae crops generally have long dormancy periods. Domesticated crops inherited these longer dormancy alleles from their wild progenitors, which have since been modified and selected during cultivation and utilization by humans. Thus, allelic combinations at different seed dormancy loci are currently represented in Triticeae germplasm preserved in seed repositories and gene banks as accessions and materials of breeding programs. Methods to evaluate seed dormancy are key to explore, analyze, and exploit optimal alleles in dormancy genes. Recent developments in genomics have accelerated the identification and analysis of seed dormancy loci in Triticeae species. Transgenic experiments have been conducted to validate if candidate genes affect seed dormancy and more recently have yielded an array of mutations derived from genome editing for practical applications. The information gathered on these seed dormancy loci provides a deeper knowledge of germplasm diversity and offers strategies to control seed dormancy in breeding programs in Triticeae crops.

Key words Barley, Wheat, Pre-harvest sprouting, Genetic resources, Genetic analysis

1 Seed Dormancy in Wild Progenitors

Dormancy can be defined as the failure of an intact, viable seed to germinate under conditions that would normally be favorable for germination [1]. Seed dormancy in wild plants is thought to reflect one of the most fundamental adaptation traits of their original habitat. Both barley and wheat (*Triticum* sp.) were domesticated ca. 10,000 years ago from their wild progenitors around the Fertile Crescent [2].

Barley (*Hordeum vulgare* ssp. *vulgare*) was domesticated from *H. vulgare* ssp. *spontaneum* [3, 4]. This wild barley usually shows a long grain dormancy period that lasts for several months after full grain maturation [5]. This long dormancy means that grains will at first not germinate in response to transient moisture availability and will therefore survive hot, dry summers. However, after several months of exposure to higher temperatures during the summer,

Naoto Kawakami and Kazuhiro Sato (eds.), *Seed Dormancy: Methods and Protocols*, Methods in Molecular Biology, vol. 2830, https://doi.org/10.1007/978-1-0716-3965-8_2,

grain dormancy is broken, which allows the grains to germinate after rains in the fall [6].

The wild progenitors of wheat share a similar habitat as wild barley. The wild progenitor (*Triticum boeoticum*: AA genome) of cultivated diploid wheat (*T. monococcum*) also has a longer seed dormancy than the cultivated form [7]. The wild progenitor of tetraploid wheat (*T. turgidum ssp. diccocoides*: AABB genome) also displays a stronger dormancy [8] than domesticated tetraploid wheat (*T. dicoccum* ssp. *durum*). However, the A genome donor species (*T. urartu*) of domesticated tetraploid wheat is different from the species that contributed the A genome for diploid cultivated wheat. Hexaploid common wheat (*T. aestivum*: AABBDD genome) evolved from the natural hybridization of domesticated tetraploid wheat and *Aegilops tauschii* (DD genome), which is also characterized by a longer seed dormancy [9]. Thus, longer seed dormancy in domesticated wheat can be traced back to the alleles harbored by their wild progenitors.

2 Selection of Seed Dormancy After Domestication

Wild progenitors of Triticeae crops grow in the winter and flower in the spring in the Middle East and require several weeks of exposure to cold temperatures for flowering, i.e., vernalization. Domesticated Triticeae crops have mutations in vernalization genes to allow them to grow in highland or higher latitude areas, where extremely long seed dormancy would preclude crop cultivation. Thus, the selection pressure for shortened seed dormancy periods has been driven by the postdomestication sowing activities of growers.

In general, cereal crop evolution has shortened the duration of seed dormancy. A classic example involves malting barley kernels for brewing beer or distilling whisky, which relies on controlled germination [6]. Pedigrees of cultivars show that malting barley from UK and Moravia spread to neighboring countries, East Asia, and the Americas during the nineteenth century. Long periods of seed dormancy became undesirable because they required a longer storage period prior to malting.

In Asian monsoon areas, the rainy season tends to overlap with the harvest season for wheat and barley in June. Both wheat and barley thus often suffer from sprouting of kernels on the spike (known as preharvest sprouting). For example, in Japan, an official report mentioning preharvest sprouting was published as early as in 711 A.D. (Nara era), about 1300 years ago. Early maturing landraces with longer seed dormancy have therefore been selected during cultivation practices to minimize preharvest sprouting.

Thus, the control of seed dormancy is of great practical importance in agricultural production of crop species and in grain utilization by food and beverage industries.

3 Methods for the Evaluation of Seed Dormancy

For genetic analysis and trait improvement, a germplasm shows a desired trait is a prerequisite. Methods to evaluate seed dormancy have been developed in laboratories and by breeding programs when they need to determine the phenotype of a germplasm of interest.

To analyze hundreds or thousands of accessions, methods for the simultaneous evaluation of the trait in multiple samples are essential. Conversely, when the trait is analyzed using genetic mapping to narrow down the region of interest containing a candidate gene, a method for evaluating individual plants is necessary. Thus, the development of several analytical methods is key to the genetic investigation and improvement of this trait.

Greater preharvest sprouting tolerance has been one of the most important target traits improved by wheat and barley breeding programs, especially in Asian monsoon areas. In Chap. 16, Fujita et al. describe examples of evaluation methods for preharvest sprouting in wheat breeding programs used in Japan. Japanese arable land covers multiple climatic growing conditions for wheat, with spring/winter accessions grown in specific locations. Each breeding program has its specific evaluation method for screening plants and/or related genetic marker analysis. The level of preharvest sprouting tolerance is related to seed dormancy, which itself is influenced by the growing conditions experienced by crops. Thus, the methods of evaluation for preharvest sprouting have attempted to focus on reliable and stable scoring methods. The same robust methods are also used in the analysis of tolerance in early hybrid generations of breeding programs.

4 Traits Associated with Seed Dormancy

There are several traits related to seed dormancy or preharvest sprouting in Triticeae crops. One example is the white/red grain color in common wheat. Wheat cultivars with white grain are generally considered to be more sensitive to preharvest sprouting than their red grain relatives. Since seed color is an important category to classify wheat grains and to determine the quality of wheat products and what flour can be used for, we cannot simply select red grained wheat for their increased tolerance to preharvest sprouting. In Chap. 4, Himi and Kurihara present how the seed color regulator gene *R-1*, encoding a MYB-type transcription factor, can be used as a tool for estimating and modulating seed dormancy over the course of seed development. As hexaploid wheat harbors three genomes, selection of a given trait by DNA markers requires special attention. The same principle holds true when isolating mutations of interest in each subgenome to establish a desired genotype by intercrossing [10] or genome editing (see also Chap. 13, Abe et al.).

5 Diversity of Dormancy in Genetic Resources

A good example of the extent of diversity for seed dormancy was provided by a gene bank–scale evaluation of barley dormancy for thousands of accessions at Okayama University [5] (*see* also Fig. 1). The level of seed dormancy was scored in 4365 barley cultivars by planting every accession in the field to harvest fresh grain samples. Domesticated barley accessions exhibited vast variation in their level of dormancy and a clear differentiation in the dormancy periods of cultivars collected from different regions. This dataset allowed the identification of several loci for dormancy genes [7], some of which have since been cloned [6, 11] (see also Chap. 10, Nakamura and Sato).

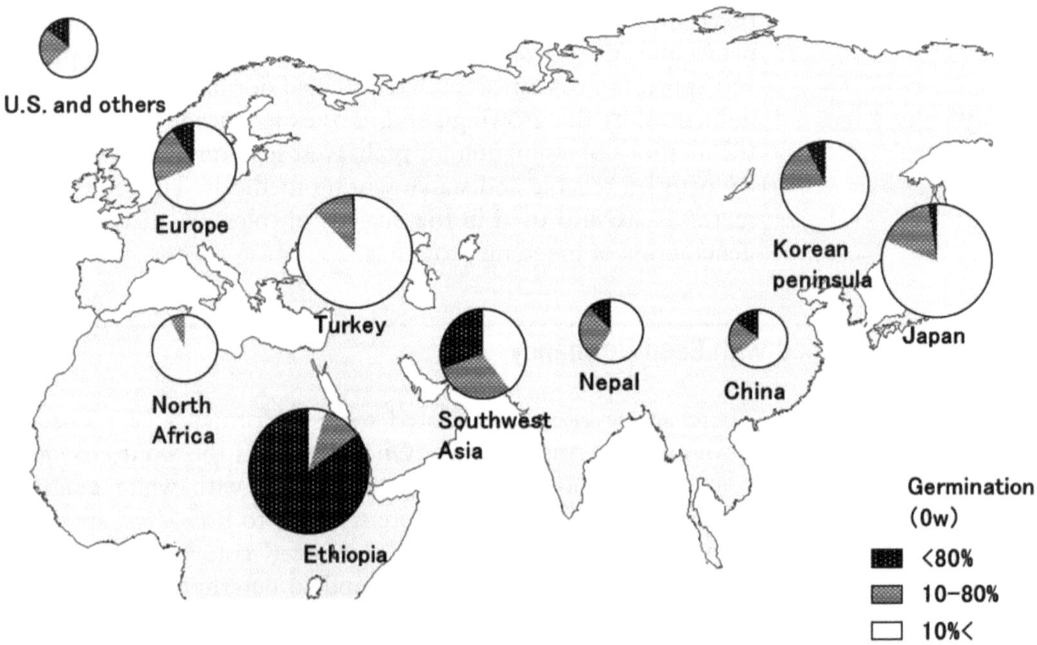

Fig. 1 Variation in seed dormancy (germination just after harvest) for a collection of cultivated barley maintained at Okayama University, Japan [5]

In Chap. 11, Yang and Wang present a genome-wide association study (GWAS) of preharvest sprouting in wheat. We hope that additional candidate loci for dormancy genes will be isolated, if and when the gene bank–scale phenotypic evaluation is combined with GWAS analysis and results.

The evaluation of breeding pedigree also provides important information to understand the standing variation in breeding germplasm. Chono et al. [14] analyzed the pedigree of Japanese wheat cultivars and traced the inheritance pattern for alleles of the wheat dormancy gene *MOTHER OF FT AND TFL1 (MFT)* [13]. In addition, several Japanese cultivars carry loss-of-function mutations in the abscisic acid (ABA) catabolism gene *TaABA8′OH1*, encoding an ABA 8′ hydroxylase, leading to higher ABA levels in seeds and thus resulting in reduced seed germination at harvest [12].

A detailed picture of seed dormancy alleles in breeding history may provide new strategies to further improve genotypes suitable for specific breeding programs and cultivation areas.

6 Genomic Resources to Identify Seed Dormancy Genes

Multiple genomic resources have been developed to analyze partial or complete genomic sequences and their associated functions in barley and wheat [15].

High-quality genetic maps are essential tools to localize genetic components on the genome and connect nucleotide sequences to Mendelian segregation patterns in genetic analysis. For barley, Sato et al. [16] developed a high-resolution genetic map of barley using polymorphisms within the transcripts of 2890 loci between the two founding parents of a single mapping population. High-throughput and high-quality multiplex PCR-based genotyping assays now allow allele-specific detection of single-nucleotide polymorphisms (SNPs) between germplasm. A consensus genetic linkage map containing over 2900 gene-based SNP markers was developed (http://harvest.ucr.edu/) [17].

Several bacterial artificial chromosome (BAC) libraries have been constructed for barley and wheat and are used as starting material for genome sequencing. A comprehensive set of full-length cDNA (FL-cDNA) sequences for barley was collected by Sato et al. [18]; another set of ca. 25,000 clones was generated and sequenced from the same cultivar across 40 organs and treatments [19]. Mochida et al. [20] generated almost 16,000 FL-cDNA sequences in the wheat cultivar Chinese Spring. These FL-cDNA sequences in both species have allowed the annotation of genes on BACs and on genome assemblies, or even in other species to identify orthologous genes.

The International Genome Sequencing Consortium has published high-quality genome sequences for barley [21, 22] and wheat [23, 24]. Recently, multiple high-quality chromosome-scale assemblies were published for wheat and barley [25, 26]. These resources may accelerate the identification of seed dormancy loci more efficiently by providing genetic markers or candidate sequences.

7 Genetic Analysis of Seed Dormancy

Most grain quality traits are quantitatively inherited and are therefore typically approached as quantitative trait loci (QTLs). Several examples of QTL linkage mapping and subsequent map-based cloning are described by Nakamura and Sato in Chap. 10.

If multiple loci control a trait, a substitution line carrying a single QTL may be used for mapping. Such an approach was followed to develop a segregating population of 910 F_2 plants to map the dormancy gene *Quantitative trait locus seed dormancy 1* (*Qsd1*) [6]. A cDNA-based genetic marker cosegregated with *Qsd1* in 4792 F_3 plants, which narrowed down the *Qsd1* region to 9467 bp between two markers.

One of the limitations of single plant–based QTL mapping is the lack of evaluation methods available to phenotype single plants. If the phenotypic effect is small, a progeny test is required to confirm the phenotype and genotype, as demonstrated for the fine-mapping of the dormancy gene *Qsd2* [11]; however, it requires much more effort for phenotyping than when assessing single plants.

However, such Mendelian genetic analyses are still an effective method to identify genes without previous knowledge of the underlying causal gene.

8 Transformation to Validate the Function of Dormancy Genes

Phenotypic changes caused by genetic transformation or genome editing offer the most direct evidence to connect the effect of a nucleotide sequence and seed dormancy. The strategies implemented to modify the genome will differ depending on whether the gene in question is present in a cultivar amenable to transformation, such as the barley cultivar Golden Promise [6] or the wheat cultivar Fielder [27], which can be transformed via Agrobacterium (*Agrobacterium tumefaciens*).

For the seed dormancy gene *Qsd1*, the Golden Promise cultivar harbors a functional allele, thus making functional complementation impossible by the introduction of a transgene expressing another allele. Another possible strategy can be undertaken, as

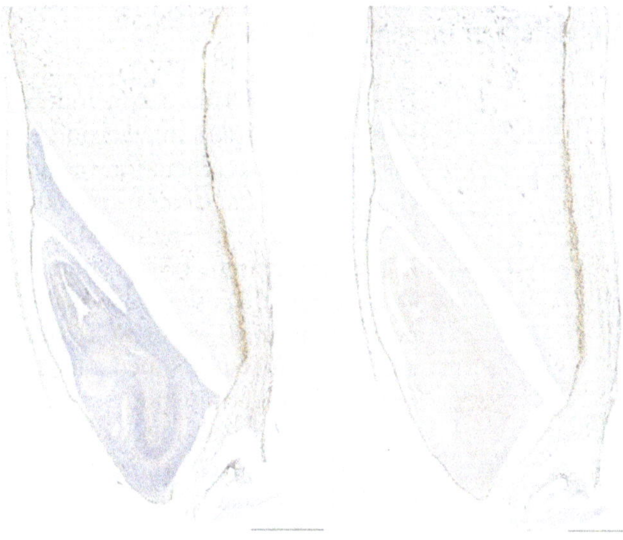

Fig. 2 Expression of *Qsd1* in dormant (*qsd1* allele) embryos 19 days after flowering, as detected by in situ hybridization with an antisense (left) or sense (right) *qsd1* probe against the 3′ untranslated region of *Qsd1* [6]

described by Sato et al. [6], who developed a *Qsd1* knockdown line using RNA interference (RNAi) to explore the effects of this gene on seed germination. They also conducted complementation tests using genetic substitution lines carrying the recessive long dormancy *qsd1* allele in the Golden Promise background and transforming them with a construct driving the expression of the cloned short dormancy *Qsd1* allele, demonstrating the effects of *Qsd1* on seed germination.

Transcriptional profiling of *Qsd1* helped define the precise developmental window during which its encoded alanine amino transferase (AlaAT) enzyme accumulated in the grain [6]. *Qsd1/qsd1* transcript levels were detected in developing grains at 21, 28, and 35 days after flowering. The grain-specific expression pattern was further analyzed using RNA in situ hybridization after flowering in the long dormancy haplotype (*qsd1*) (Fig. 2), revealing the accumulation of transcripts specifically in the embryo but not in the endosperm.

Transient assays may also provide another means to estimate the contribution of a given gene to dormancy. In Chap. 12, Abe et al. describe a protocol for transiently expressing genes related to seed dormancy in the scutellum of immature wheat embryos to analyze their effects on germination.

9 Modification of Dormancy Alleles

Knockout mutations can be induced by targeted genome modification (known as genome editing), which offers a powerful method for the efficient generation of mutations in desired genes. Hisano et al. (Chap. 14) demonstrate that both *Qsd1* and *Qsd2* are dormancy genes by generating knockout mutations in both genes and in the same genetic background via genome editing. Likewise, Abe et al. (Chap. 13) used the available information on the barley *Qsd1* gene to modify three orthologous genes in the wheat subgenomes and successfully developed a triple knockout mutant with a significant delay in germination.

The barley genome has been previously edited using clustered regularly interspaced short palindromic repeats (CRISPR) and the CRISPR-associated nuclease Cas9; however, transformation and genome editing experiments may suffer from limitations stemming from the low transformation efficiency of some accessions. Indeed, the cultivar Golden Promise is one of the few haplotypes that can be reliably transformed via Agrobacterium. Notably, Hisano and Sato [28] identified loci controlling transformation potential mapping to chromosomes 2H and 3H, using an F_2 population derived from a cross between the cultivar Golden Promise and the transformation-recalcitrant cultivar Haruna Nijo. Introducing these genomic regions into a desired haplotype may increase its transformation efficiency and thus its genome editing capabilities for grain quality traits.

A genotype-independent genome editing system would have great advantages for the direct phenotypic confirmation of a target gene or allele. Luo et al. (Chap. 15) demonstrate the application of genome editing at *TaQsd1* in commercial wheat cultivars.

These techniques may accelerate the generation of desirable haplotypes suitable for agricultural production in countries where a genome-edited cultivar is permitted for cultivation, such as in Japan.

10 Application of Seed Dormancy Loci to Prevent Preharvest Sprouting

The cloning of seed dormancy loci would open new avenues to improving tolerance to preharvest sprouting using marker-assisted selection. As illustrated with the genes *MFT* and *Mitogen-activated Protein Kinase Kinase 3* (*MKK3*), we determined the distribution of the causal sequence variation in seed dormancy loci in worldwide cultivars. Thus, we are now in a position to introduce dormant alleles into a given cultivar by marker-assisted selection to increase its tolerance to preharvest sprouting. The dormant allele of *MFT* is predominately limited to wheat cultivars from the southern part of

Japan. We may thus hope that introducing the dormant *MFT* allele into cultivars from the northern part of Japan or in other geographic regions will raise their tolerance. In fact, such cases have already been reported in Japan and China.

Aside from marker-assisted selection, information regarding the identified gene can be used to generate a new loss-of-function allele using genome editing. Some results along these lines have been reported in wheat and barley using *AlaAT* and *MKK3* by several research groups in Japan [29, 30].

Appropriate seed dormancy strength differs depending on the intended use of the cultivar, such as feeding or malting in barley, and depending on the growing region. Generally, the more the growing region extends north, the shorter the time between harvest and sowing will be, thus necessitating lower dormancy levels for uniform germination. We can apply our knowledge of seed dormancy loci to develop cultivars with appropriate levels of seed dormancy by selecting a combination of these loci to realize tailor-made breeding for tolerance to preharvest sprouting.

11 Future Perspectives

Natural and induced variations provide opportunities to analyze grain traits; however, combining genotypes and phenotypes remains challenging. To estimate the sequence diversity of barley germplasm, the Leibniz Institute of Plant Genetics and Crop Plant Research (IPK) in Gatersleben, Germany, genotyped single plants from 21,405 accessions in their crop gene bank via genotype-by-sequencing (GBS) [31]. All genome reads were aligned to the reference genome from the barley cultivar "Morex," which allowed the detection of 171,263 biallelic SNPs. The geographical origins of each sample were consistent with the SNP diversity analysis. Using this massive amount of SNP data, loci for some traits were precisely mapped. Thus, genotyping an entire gene bank is one of the solutions for the rapid identification and isolation of loci responsible for grain quality traits.

We also realize the need to establish genetically stable reference accessions to obtain consistent results for genome and proteome sequencing [32, 33]. It has become clear that the parallel sequencing of multiple haplotypes is essential for enhancing our understanding of the genetic and genomic features underlying natural and induced variations in various traits. Representative accessions based on the IPK GBS dataset have been selected to analyze the barley pan-genome [34], which may represent a better and more comprehensive complement of sequence variation within the barley genome and can be used to identify target gene sequences regulating grain quality.

We were incredibly encouraged when the wheat genome was edited based on the information gathered from barley *Qsd1* to produce a triple mutant with modified seed dormancy. The catalog of dormancy genes in other plant species will be a resource of variation to modulate seed dormancy in Triticeae crops.

The improvement of tolerance to preharvest sprouting requires more detailed knowledge on the type of variation affecting seed dormancy. The identification of other alleles for known seed dormancy genes, such as *AlaAT*, *MKK3*, and *MFT*, and of the causal genes for mapped but yet to be cloned seed dormancy QTLs is necessary. The successful identification of currently known seed dormancy genes has demonstrated its feasibility for identifying other causal genes for seed dormancy QTLs and indicates that it is possible to actively conduct such research around the world. In addition, the identification of the causal genes will make it possible to knock them out via genome editing methods to create new alleles and germplasm for seed dormancy. Recent advances in genomic analysis and bioinformatics will facilitate our progress in this undertaking. Just as the accurate prediction of heading date has progressed, we envisage that an analysis of the relationships between natural or artificial variation in seed dormancy genes and their effect on seed dormancy using artificial intelligence might allow us to predict, at least to some extent, the response of breeding lines for seed dormancy as well as we can currently predict the effects associated with qualitative monogenic traits such as seed coat color, disease resistance, or amylose contents in flour.

Due to the effects of global warming, crop growing conditions are becoming more instable, especially with the increase of temperature and precipitation. Cereal breeding programs must pay more attention to changes in growing conditions and improve tolerance to preharvest sprouting accordingly.

References

1. Bewley JD (1997) Seed germination and dormancy. Plant Cell 9:1055–1066
2. Zohary D, Hoph M, Weiss E (2012) Domestication of plants in the Old World, 4th edn. Oxford University Press, New York, pp 20–74
3. von Bothmer R, Sato K, Komatsuda T et al (2003) The domestication of cultivated barley. In: von Bothmer R, van Hintum T, Knüpffer H, Sato K (eds) Diversity in barley (*Hordeum vulgare*). Elsevier Science B.V, Amsterdam, pp 9–27
4. Pourkheirandish M, Hensel G, Kilian B et al (2015) Evolution of the seed dispersal system in barley. Cell 162:527–539
5. Takeda K, Hori K (2007) Geographical differentiation and diallel analysis of seed dormancy in barley. Euphytica 153:249–256
6. Sato K, Yamane M, Yamaji N et al (2016) Alanine aminotransferase controls seed dormancy in barley. Nat Commun 7:11625
7. Hori K, Sato K, Takeda K (2007) Detection of seed dormancy QTL in multiple mapping populations derived from crosses involving novel barley germplasm. Theor Appl Genet 115:869–876
8. Ohta S (2022) Genetic variation and genetic control of intraspikelet differences in grain weight and seed dormancy in wild and domesticated emmer wheats. Breed Sci 72:198–212

9. Ahmed MIY, Gorafi YSA, Kamal NM et al (2023) Mining Aegilops tauschii genetic diversity in the background of bread wheat revealed a novel QTL for seed dormancy. Front Plant Sci. https://doi.org/10.3389/fpls.2023.1270925
10. Nakamura T, Yamamori M, Hirano H et al (1995) Production of waxy (amylose-free) wheats. Mol Gen Genet 248:253–259
11. Nakamura S, Pourkheirandish M, Morishige H et al (2016) Mitogen-activated protein kinase kinase 3 regulates grain dormancy in barley. Curr Biol 26:775–781
12. Chono M, Matsunaka H, Seki M et al (2013) Isolation of a wheat (*Triticum aestivum* L.) mutant in ABA 8′-hydroxylase gene: effect of reduced ABA catabolism on germination inhibition under field condition. Breed Sci 63:104–115
13. Nakamura S, Abe F, Kawahigashi H et al (2011) A wheat homolog of *MOTHER OF FT AND TFL1* acts in the regulation of germination. Plant Cell 23:3215–3229
14. Chono M, Matsunaka H, Seki M et al (2015) Molecular and genealogical analysis of grain dormancy in Japanese wheat varieties, with specific focus on *MOTHER OF FT AND TFL1* on chromosome 3A. Breed Sci 65:103–109
15. Sato K (2020) History and future perspectives of barley genomics. DNA Res 27. https://doi.org/10.1093/dnares/dsaa023
16. Sato K, Nankaku N, Takeda K (2009) A high-density transcript linkage map of barley derived from a single population. Heredity 103:110–117
17. Close TJ, Bhat PR, Lonardi S et al (2009) Development and implementation of high-throughput SNP genotyping in barley. BMC Genomics 10:582–594
18. Sato K, Shin-I T, Seki M et al (2009) Development of 5006 full-length cDNAs in barley: a tool for accessing cereal genomics resources. DNA Res 16:81–89
19. Matsumoto T, Tanaka T, Sakai H et al (2012) Construction and sequence analysis of barley 24,783 full-length cDNAs. Plant Physiol 156:20–28
20. Mochida K, Yoshida T, Sakurai T et al (2009) TriFLDB: a database of clustered full-length coding sequences from Triticeae with applications to comparative grass genomics. Plant Physiol 150:1135–1146
21. International Barley Genome Sequencing Consortium (2012) A physical, genetic and functional sequence assembly of the barley genome. Nature 491:711–716
22. Mascher M, Gundlach H, Himmelbach A et al (2017) A chromosome conformation capture ordered sequence of the barley genome. Nature 544:427–433
23. International Wheat Genome Sequencing Consortium (IWGSC) (2014) A chromosome-based draft sequence of the hexaploid bread wheat (*Triticum aestivum*) genome. Science 345:1251788
24. IWGSC (2018) Shifting the limits in wheat research and breeding using a fully annotated reference genome. Science 361:eaar7191
25. Jayakodi M, Padmarasu S, Haberer G et al (2020) The barley pan-genome reveals the hidden legacy of mutation breeding. Nature 588:284–289
26. Walkowiak S, Gao L, Monat C et al (2020) Multiple wheat genomes reveal global variation in modern breeding. Nature 588:277–283
27. Ishida Y, Tsunashima M, Hiei Y, Komari T (2015) Wheat (*Triticum aestivum* L.) transformation using immature embryos. Methods Mol Biol 1223:189–198
28. Hisano H, Sato K (2016) Genomic regions responsible for amenability to Agrobacterium-mediated transformation in barley. Sci Rep 6:37505
29. Abe F, Haque E, Hisano H, Tanaka T, Kamiya Y, Mikami M, Kawaura K, Endo M, Onishi K, Hayashi T, Sato K (2019) Genome-edited triple-recessive mutation alters seed dormancy in wheat. Cell Rep 28:1362–1369
30. Hisano H, Hoffie RE, Abe F, Munemori H, Matsuura T, Endo M, Mikami M, Nakamura S, Kumlehn J, Sato K (2021) Regulation of germination by targeted mutagenesis of grain dormancy genes in barley. Plant Biotechnol. https://doi.org/10.1111/pbi.13692
31. Milner SG, Jost M, Taketa S et al (2019) Genebank genomics highlights the diversity of a global barley collection. Nat Genet 51:319–326
32. Schreiber M, Mascher M, Wright J et al (2020) A genome assembly of the barley 'transformation reference' cultivar golden promise. G3 10:1823–1827
33. Sato K, Abe F, Mascher M et al (2021) Chromosome-scale genome assembly of the transformation-amenable common wheat cultivar 'Fielder'. DNA Res 29:dsab008
34. Jayakodi M, Schreiber M, Stein N et al (2021) Building pan-genome infrastructures for crop plants and their use in association genetics. DNA Res 28:dsaa030

Part II

Dormancy, Germination and Longevity Assays

Chapter 3

Assessment of Dormancy Level and Germination Ability of Seeds with Physiological Dormancy

Lipeng Zheng, Masahiko Otani, and Naoto Kawakami

Abstract

Germination test is fundamental and commonly used technique for seed dormancy and germination studies, and proper assessment of dormancy level and germination ability of a given set of seeds is prerequisite for most of the studies. However, germination is very sensitive to imbibition conditions, and dormancy development is also sensitive to growth conditions of the mother plants. In this chapter, we describe tips for plant growth and germination test mainly for physiological and molecular genetic studies with Arabidopsis. This protocol can be applied for other plant species with relatively small seeds and for various studies to analyze the effect of light, phytohormones, and other chemicals in seed germination.

Key words Arabidopsis, Physiological dormancy, Germination, Temperature

1 Introduction

Seed dormancy has been classified into several types, and physiological dormancy is most commonly observed among weed and crop species [1]. The seeds with morphological dormancy have underdeveloped embryos, but there are no visible signs of dormancy levels in the seeds with physiological dormancy. In seed dormancy and germination studies, we need to assess dormancy levels of a given set of seeds by analyzing and comparing germination of the seed populations. However, dormancy development has been shown to be sensitive to growth conditions of mother plants such as light, temperature, soil moisture, and chemical components [2]. We often get different germination data from different seed batches even if the plants were grown in the same conditions. The fluctuation between seed batches may be caused by position of plant(s) in a growth chamber, frequency of watering, nutrient solutions, humidity, seed harvest timing, and other unidentified differences between experiments. Germination is also sensitive to imbibition conditions and time after harvest. Germination data

Naoto Kawakami and Kazuhiro Sato (eds.), *Seed Dormancy: Methods and Protocols*, Methods in Molecular Biology, vol. 2830,
https://doi.org/10.1007/978-1-0716-3965-8_3,

obtained from the same batch of seeds that were imbibed with pure water and some nutrient solutions must be different since nitrate breaks dormancy and stimulates germination. Seeds are also sensitive to imbibition temperature, substrate moisture, chemical compounds in the solution, aeration, light intensity, and quality.

It has been shown that seed responsiveness to temperature is closely related to the level of dormancy in soil-buried seeds of winter and summer annuals [1]. In the case of winter annuals, the maximum temperature for germination rises gradually during an after-ripening period in the summer, but germination is repressed by environmental temperatures higher than the upper limit for germination. The seeds become germinable in high-temperature conditions in autumn, and germinate in the field when the temperature falls below the upper limit for germination [3]. This temperature sensitivity change during after-ripening is also observed in the seeds produced in growth chamber and stored in laboratory conditions [4]. Interestingly, the seeds of reduced dormancy mutants *rdo1, rdo2, rdo3,* and *rdo4* showed germination at supraoptimal temperatures, and the seeds of high-temperature-resistant germination mutants *trg1, trg2, trg3, trg4/tt7-4,* and *trg5/abi3-14* showed reduced dormancy phenotype [4]. Similar correlation between dormancy and the seed sensitivity to temperature has also been reported in several wheat cultivars and lines [5, 6]. The genetic linkage between dormancy and the responsiveness to temperature, and the physiological correlation between after-ripening and the responsiveness to temperature suggest that a temperature sensing and responding mechanism in the seeds is essential for dormancy of seeds with physiological dormancy.

In most of the seed dormancy studies, dormancy levels of the seeds have been evaluated by germination at one specific temperature condition, usually at the optimum temperature for germination. However, even when the two seed sets show 100% germination, the dormancy levels may different between the two. For example, freshly matured and after-ripened seeds of Arabidopsis Col-0 accession seeds showed almost the same germination speed and final percentage (100%) germination at 15 °C, but they showed apparently different germination percentages at 26 °C (Fig. 1). It means that dormancy levels can be assessed in high resolution by testing germination at different temperature conditions.

In this chapter, we describe more accurate protocol for evaluating seed dormancy levels of Arabidopsis. This protocol can be modified for other plant species such as lettuce, spinach, and cabbage by selecting plate size and/or using imbibition substrate (e.g., filter paper). This protocol can also be used for germination studies to analyze the effects of light, phytohormones, some inhibitors, and other chemicals. General guide for seed dormancy and germination study has been well summarized [7].

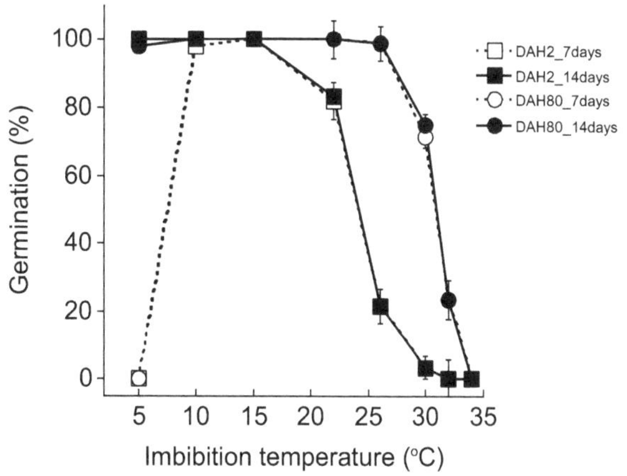

Fig. 1 Germination percentages of Arabidopsis Col-0 ecotype seeds after 7 and 14 days of imbibition. Freshly harvested (2 days after harvest; DAH2) and after-ripened (DAH80) seeds were imbibed as described in this protocol

2 Materials

2.1 Growing, Flowering Check, and Seed Collection

1. Micropipette with 10 μL tip.
2. 0.1% low melting agarose.
3. Soil (e.g., N: 150 mg/L, P: 600 mg/L, K: 100 mg/L).
4. Thread.
5. Tea strainer.
6. Silica gel.
7. Tightly sealed box.

2.2 Seed Sterilization

1. 70% ethanol.
2. Ethanol (ca. 99.5%).
3. Vortex mixer.
4. Clean bench.
5. Micropipette with 200 μL tip.
6. Micropipette with 1 mL tip.

2.3 Seed Imbibition

1. 24-well microplate.
2. Micropipette with 1 mL tip.
3. Plastic tape.
4. Incubator with illumination.
5. Stereo microscope.

3 Methods

3.1 Growing Arabidopsis Plants in a Growth Chamber

1. Disperse seeds in 0.5–1 mL of 0.1% low melting agarose solution in a microtube and imbibe at 4 °C for 4 days to facilitate rapid and uniform germination (*see* **Note 1**).
2. Pick the seed one by one from the agarose solution by sucking with micropipette attached with a 10 μL micro tip and put on the surface of the soil in a cup (*see* **Note 2**). Sow three seeds in a cup. Put the cups in a plastic vat in a growth chamber under continuous illumination at 22 °C (*see* **Note 3**). To avoid drying, wrap the vat with plastic wrap for a few days until green cotyledons appear.
3. Grow the plants under continuous illumination at 22 °C by rotating plant positions once in a day to equalize growth conditions. Thin out abnormal seedlings at around 2 weeks after germination, and keep one healthy seedling to grow in a cup. Supply the plants with tap water when surface of the soil become dry (*see* **Note 3**).
4. Harvest seeds from the plants at physiological maturity when about half of the fruits on a plant turned to yellow by hand threshing. Remove dry siliques, stamen, sepals, petals, and other dust by using tea strainer. To obtain seeds at specific maturation stage, tag flowers at anthesis with threads of different colors depending on the day of the week (avoid inconspicuous color such as green). Arabidopsis seeds will mature at around 24 days after flowering. Dry the seeds in a tightly sealed box with silica gel or in a desiccator for 2 days to adjust the moisture contents of the seeds before imbibition (*see* **Note 4**). Store the seeds in a tightly sealed box with silica gel or desiccator at room temperature for after-ripening. Otherwise, store the seeds at −80 °C in a tightly sealed box with silica gel to maintain dormancy (*see* **Note 5**).

3.2 Germination Test at Multiple Temperature Conditions

1. Put the seeds into a 1.5 mL microtube, add 1 mL of 70% ethanol, and vortex or mix by inversion every minute for 5 min (*see* **Note 6**). Discard 70% ethanol by micropipette.
2. Remove moisture by washing the seeds with 1 mL of ethanol (ca. 99.5%) twice. Remove ethanol as completely as possible by inserting 200 μL pipette tip at the bottom of the tube. To facilitate evaporation of ethanol, flip the tube to disperse the wet seeds, and leave the tube in a clean bench to dry up (*see* **Note 7**).
3. Add 300 μL of ultra-pure water in a well of 24-well plate, and put 30 seeds to a well (*see* **Note 8**).

4. Pour appropriate amount of deionized water between wells, put the lid on a plate, and seal with plastic tape to avoid evaporation during incubation, especially at high-temperature conditions.
5. Imbibe the seeds at optimal, sub- and supraoptimal temperatures (e.g., 10, 25, and 32 °C) under continuous fluorescent light (18 μmol m^{-2} s^{-1}) for 7 to 14 days (*see* **Note 9**) without any stratification treatment. Count germinated seeds every 24 h by using stereo microscope (*see* **Note 10**). Germination should be scored as radicle protrusion by using stereo microscope. All germination tests should done at least in three independent seed batches with three replicates in each.
6. It is important to know viability of a seed set when the seeds did not show almost 100% germination in any temperature conditions used. Viability of the seeds can be determined by cold stratification (imbibe at 4 °C for 4 days), alternating temperature treatment (diurnal temperature fluctuation such as 25 °C/15 °C), gibberellin application, or by staining with tetrazolium solution [7].

4 Notes

1. Arabidopsis seeds do not go to the bottom and disperse in 0.1% low melting agarose solution. The seeds in a small tube can be stratified in a refrigerator. The cold treatment after seed sowing is also effective.
2. Soils for raising seedling that contain peat moss, vermiculite, and appropriate amount of nutrients are compatible for Arabidopsis growing. Growing medium (e.g., OasisR, H-1.5, Smithers-Oasis, Ohio, USA) can also be used to grow the plants with nutrient solution [8].
3. Commonly used Arabidopsis ecotypes, Col-0, L*er*, and Ws, do not require vernalization, and set seeds after 1.5–2 months after sowing in the continuous light condition. Long-day conditions such as 16/8 h light/dark cycle are also good for Arabidopsis. All the growing conditions, such as soil, nutrient supply, watering, and irradiation, should be as constant as possible.
4. This drying step can be skipped when the seeds are harvested at specific maturation stage by tagging flowers at anthesis with threads.
5. The freezing restrains after-ripening of the seeds, but dormancy still slowly decreased in the seeds even at −80 °C. We recommend to perform germination assay of the frozen dormant seeds within 6 months.

6. One thousand Arabidopsis seeds are usually 20 mg. Up to 50 mg seeds can be sterilized in a tube. The surface sterilization with ethanol is a mild method and does not so much affect dormancy levels of the treated seeds. However, in the case of no contamination of fungi or bacteria, we recommend to use seeds without any sterilization. In contrast, if you have heavy contamination, the seeds can be sterilized with aqueous solution containing 2% sodium hypochlorite and 0.1% Triton X-100. In this case, the seeds should be washed with autoclaved pure water for at least five times. Please keep sterilization time as short as possible since sodium hypochlorite solution sometimes reduces dormancy levels of the seeds.
7. The remaining ethanol absorbs moisture in the air, and the seeds imbibe the water during storage, and it reduces seed viability. Vacuum drying is also applicable in this step.
8. Over 300 μL of water in a well of 24-well plate may be anaerobic and inhibit germination. Thirty seeds in a well of 24-well plate will facilitate discrimination of germinated and ungerminated seeds. Twelve-well plate should be fine for 50 seeds in a well.

   Aqueous solution of a specific chemical such as abscisic acid or some inhibitor can also be used in this method. The hydrophobic chemicals should be first dissolved in amphiphilic solvents such as dimethylsulfoxide (DMSO) or ethanol, and then diluted to the final concentrations with ultra-pure water. Final concentration of DMSO should be lower than 0.1% since DMSO affects germination of Arabidopsis seeds at higher concentrations. Please note that control of the chemical treatment should be DMSO solution with the same concentration as chemical solutions.
9. Imbibition temperatures and a period of imbibition should be determined by preliminary experiments with wide range of temperature. Arabidopsis seed germination is progressively inhibited at supraoptimal high temperatures, and slowed down at suboptimal temperatures (Fig. 2). Germination at supraoptimal temperatures is different between ecotypes. L*er* and Ws seeds show higher and lower percentage germination than Col-0, respectively. Incubators for the multiple-temperature germination test should be the same type, or use an incubator with multiple small chambers. A thermogradient table can also be used. We recommend to test germination at least at three different temperatures to evaluate dormancy levels or investigate seed sensitivity to temperature.
10. Germination time course is informative, but germination index (GI) is often used to take speed of germination into consideration. A GI is calculated with maximum weight given to the

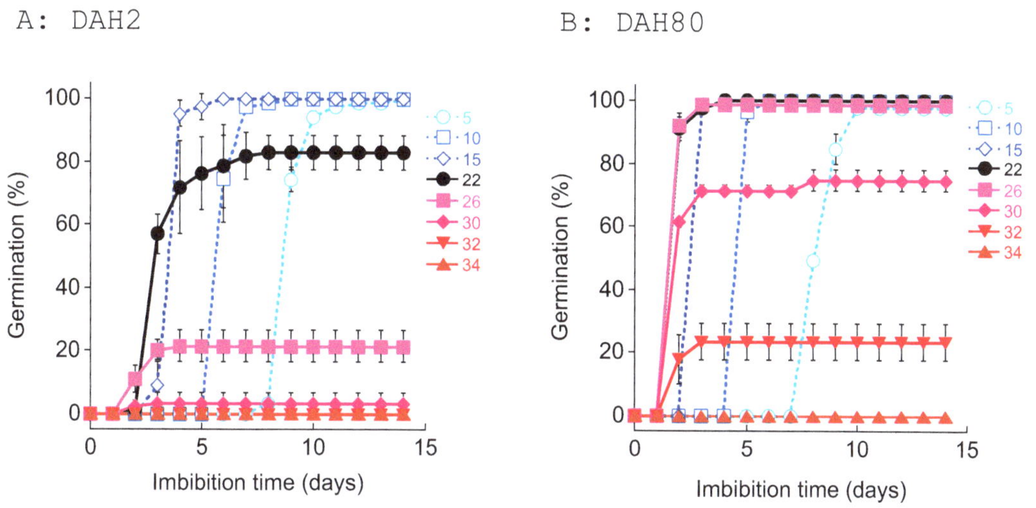

Fig. 2 Germination time course of Arabidopsis Col-0 ecotype seeds at various temperature conditions. (**a**) Freshly harvested (2 days after harvest; DAH2) seeds. (**b**). After-ripened (DAH80) seeds

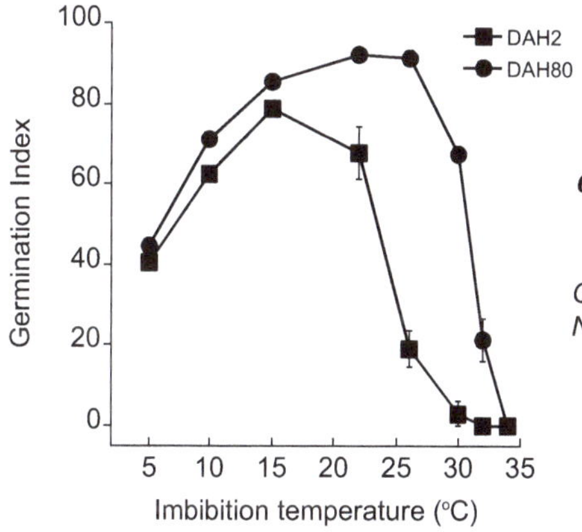

$$GI = \sum_{n=1}^{N} \frac{G_n(N+1-n)}{N}$$

Gn : percentage of germinated seeds at day *n*
N : total imbibition time (days)

Fig. 3 Germination index (GI) of Arabidopsis Col-0 ecotype seeds. Freshly harvested (2 days after harvest; DAH2) and after-ripened (DAH80) seeds were imbibed for 14 days as described in this protocol. GI was calculated as follows: $GI = (14G_1 + 13G_2 + 12G_3 + \ldots + 1G_{14})/14$

seeds that germinated first and less weight to those that germinated later (Fig. 3). Several other formulas have been developed for GI.

References

1. Baskin JM, Baskin CC (2014) Seeds. Ecology, biogeography, and evolution of dormancy and germination, 2nd edn. Academic Press, San Diego
2. Bewley JD, Bradford KJ, Hilhorst HWM, Nonogaki H (2013) Seeds. Physiology of development, germination and dormancy, 3rd edn. Springer, New York
3. Baskin JM, Baskin CC (1983) Seasonal-changes in the germination responses of buried seeds of *Arabidopsis thaliana* and ecological interpretation. Bot Gaz 144:540–543

4. Tamura N, Yoshida T, Tanaka A, Sasaki R, Bando A, Toh S, Lepiniec L, Kawakami N (2006) Isolation and characterization of high temperature-resistant germination mutants of *Arabidopsis thaliana*. Plant Cell Physiol 47: 1081–1094
5. Mares DJ (1984) Temperature-dependence of germinability of wheat (*Triticum aestivum* L.) grain in relation to pre-harvest sprouting. Aust J Agric Res 35:115–128
6. Kashiwakura Y, Kobayashi D, Jikumaru Y, Takebayashi Y, Nambara E, Seo M, Kamiya Y, Kushiro T, Kawakami N (2016) Highly sprouting-tolerant wheat grain exhibits extreme dormancy and cold imbibition-resistant accumulation of abscisic acid. Plant Cell Physiol 57: 715–732
7. Hilhorst HWM (2011) Standardizing seed dormancy research. In: Kermode AR (ed) Seed dormancy: methods and protocols, Methods in molecular biology, vol 773. Humana, Totowa, pp 43–52
8. Fujiwara T, Hirai MY, Chino M, Komeda Y, Naito S (1992) Effects of sulfur nutrition on expression of the soybean seed storage protein genes in transgenic petunia. Plant Physiol 99: 263–268

Chapter 4

Germination Assays for Comparable Seed Dormancy Depth and Distinction of Seed Color by Multiplex PCR in Wheat

Eiko Himi and Shiho Kurihara-Yonemoto

Abstract

Seed dormancy is an important trait in cereal breeding, as it prevents preharvest sprouting (PHS). Although seed dormancy is a multifactorial trait, seed color has been demonstrated to be a major dormancy-related factor controlled by few genes. The *R-1* gene is a seed color regulator that encodes a MYB-type transcription factor in wheat. A set of genetic markers designed against *R-1* can provide a powerful tool for swift wheat breeding. Depth of seed dormancy varies not only among lines but also during seed development in each line. In this chapter, we describe how developmental seeds can be collected to perform germination tests, how seed color can be observed after NaOH staining, and how to genotype wheat *R-1* genes using multiplex PCR.

Key words Seed color, Seed dormancy, *Tamyb10*, Wheat

1 Introduction

Wheat (*Triticum aestivum*) seeds can be classified as red or white. Reddish pigments, known as proanthocyanidins, accumulate in the testa (seed coat) only in red wheat lines. Since a portion of the testa is turned into flour through milling, seed color will affect the color of the final products derived from wheat flour, even if filtered, and thus constitutes a very important agronomic trait. To help distinguish wheat lines based on their seed coat color, a method was developed to enhance seed color using sodium hydroxide (NaOH) [1, 2].

Seed color has long been used as a tractable genetic marker because of the easy access to and scorability of this trait. Biffen [3] discovered that the red coloration of seeds is inherited as a dominant allele named the *Red* (*R*) gene [3], which was later renamed *R-1* [4]. Three copies of *R-1* genes are harbored by the hexaploid wheat genome and are located on the long arm of chromosomes 3A, 3B, and 3D; each copy is designated *R-A1*, *R-B1*, and *R-D1*,

Naoto Kawakami and Kazuhiro Sato (eds.), *Seed Dormancy: Methods and Protocols*, Methods in Molecular Biology, vol. 2830,
https://doi.org/10.1007/978-1-0716-3965-8_4,

respectively. Similarly, white-seeded lines are homozygous for the recessive alleles *R-A1a*, *R-B1a*, and *R-D1a*. The dominant *R-1* alleles *R-A1b*, *R-B1b*, and *R-D1b* are considered to be functionally equivalent. The *R-1* gene was cloned in 2011 and shown to encode a MYB-type transcription factor, Tamyb10, which regulates the expression of proanthocyanidin biosynthetic genes such as *Chalcone synthase* (*CHS*), *Chalcone isomerase* (*CHI*), *Flavanone 3-hydroxylase* (*F3H*), and *Dihydroflavonol 4-Reductase* (*DFR*) [5, 6]. The genomic sequences of dominant and recessive alleles of *R-A1*, *R-B1*, and *R-D1* were determined, allowing the development of a set of genetic markers [5]. Although the first reported markers developed were rather complex, as they required setting up five separate reactions, a convenient multiplex PCR assay for the *Tamyb10* genotype was recently developed [7].

An association between seed color and dormancy was initially reported by Nilsson-Ehle [8] and confirmed in subsequent studies [8, 9]. Weak seed dormancy causes germination before harvest, also called preharvest sprouting (PHS). Cultivars with strong dormancy are needed worldwide to prevent damage to grain yield resulting from PHS.

The depth of seed dormancy varies over the course of seed development and maturation. While immature seeds do not germinate, premature seeds from low dormancy lines do. Moreover, the exact timing of seed developmental stages differs from year to year. As seed water content reflects their developmental stage, measuring seed water content is necessary for dormancy tests to compare seed dormancy from wheat lines collected across multiple growing seasons and years. The Germination Index can also be used to compare dormancy depth between seeds harvested from different lines on separate years [10].

In this chapter, we introduce methods for the accurate evaluation of the depth of wheat seed dormancy and for the biochemical and genetical distinction of seed color:

1. Seed sampling: seeds can be collected at the same developmental stage by tracking flowering time and by measuring seed water contents by baking.
2. Evaluation of seed dormancy depth: dormancy levels are compared by germination tests under fixed conditions and with proper calculation.
3. Seed color distinction: NaOH staining of seed enhances seed color for convenient visualization, and multiplex PCR can genotype the multiple alleles of *Tamyb10*, the major seed color determinant.

2 Materials

2.1 Seed Sampling and Measurement of Water Contents

1. Seven-colored vinyl tape (*see* **Note 1**).
2. Aluminum weighing dishes.
3. Electric balance with a precision of 1 mg.

2.2 Evaluation of Seed Dormancy Depth

1. Petri dishes, 9-cm diameter.
2. Filter paper, 84-mm diameter (*see* **Note 2**).
3. Plastic container box.
4. 10 μM abscisic acid (ABA) solution (optional) (*see* **Note 3**): Weigh 26.43 mg ABA and add 1 mL ethanol. Dissolve to make a 100 mM ABA solution and transfer to a 1.5-mL tube. Add 9.9 mL water to a 50-mL beaker with a magnetic stir bar. Add 100 μL of 100 mM ABA to the beaker drop by drop while stirring. Filter the solution with a 0.22-μm disk filter into a 1-L cylinder. Adjust volume to 1 L with water. Store at 4 °C in a bottle wrapped with aluminum foil to avoid light.
5. Razor blade (optional) (*see* **Note 4**).

2.3 Seed Color Distinction

1. 1 N NaOH. Weigh 40.0 g NaOH (*see* **Note 5**). Transfer NaOH to a glass beaker containing 900 mL water and dissolve NaOH thoroughly. Adjust volume to 1 L with distilled water (*see* **Note 6**).
2. Optional: 100 μL Tween 20, a surfactant that may be added to 1 L of 1 N NaOH (*see* **Note 7**).
3. Transparent test tubes or a 24-well plate.

2.4 Tamyb10 Genotyping

1. Primers for *Tamyb10-A1*:
   Tamyb10-LP6: 5′-CTATGTGGATGGCCTTGGAT-3′
   Tamyb10-RP4: 5′-CCTGACGATGAGCTCCTCTT-3′
   Tamyb10-LP7: 5′-TTTCAATCGAGTGGGCATAA-3′
   Tamyb10-LP8: 5′-TCCCTACATGGGAGACAGAGA-3′
   ARP1: 5′-CCGCAGCATTTAACTTGTAGC-3′
2. Primers for *Tamyb10-B1*:
   Tamyb10-LP2: 5′-AGCAAGAGGAACCTGCAGTC-3′
   Tamyb10-RP1: 5′-GATGCCCTCCAGATCAAGGT-3′
3. Primers for *Tamyb10-D1*:
   DLP7: 5′-CATCGAGCTGGGCTTCAT-3′
   DRP7: 5′-TCTGGCATTTTGACAGTTCC-3′
4. Wheat genomic DNA (*see* **Note 8**).

5. DNA polymerase: *TaKaRa Ex Taq* Hot Start Version supplied by Takara Bio Inc. (*see* **Note 9**).
6. 50% (v/v) glycerol. Measure 50 mL glycerol in a 50-mL centrifuge tube and transfer the glycerol to a 200-mL beaker (some glycerol will remain on the inner wall surface of the tube). Add about 10 mL of water to the tube and screw the lid closed; then, shake vigorously to homogenize the glycerol and water. Transfer the liquid to the beaker. Rinse the tube with 10 mL of water twice and mix these solutions thoroughly with a magnetic stirrer. Adjust volume to 100 mL and transfer the mixture to a glass bottle. Autoclave the mixture. Store 50% glycerol at room temperature.
7. Electrophoresis-grade agarose.
8. 10× Electrophoresis buffer (Tris acetate EDTA [TAE] buffer): Weigh 48.4 g Tris–HCl and transfer to a beaker. Add 900 mL water, 11.4 mL glacial acetic acid, and 20 mL of 0.5 M EDTA solution (pH 8.0). Mix and adjust volume to 1 L with water. Dilute stock solution 10:1 using ultrapure water to make a 1× working solution.

3 Methods

3.1 Seed Sampling

1. Assign a different vinyl tape color for each day of the week: for example, Sunday, red; Monday, white; Tuesday, orange; Wednesday, gray; Thursday, green; Friday, blue; and Saturday, yellow.
2. Tag spikes on flowering day with a corresponding colored vinyl tape (*see* **Note 10**).
3. Write the week number on the tag (*see* **Note 11** and Fig. 1a, b).
4. Harvest spikes on a desirable day and collect seeds from only the first and second florets of the central spikelets on spikes (*see* **Note 12** and Figs. 2 and 3).

3.2 Measurement of Seed Water Contents

1. Count seeds for measurement of water contents (*see* **Note 13**).
2. Weigh an empty aluminum dish.
3. Weigh seeds on the aluminum dish and calculate fresh seed weight.
4. Bake the seeds in the dish at 150 °C for 3 h.
5. Weigh the seeds again on the aluminum dish (*see* **Note 14**).
6. Discard the baked seeds, and only weigh the aluminum dish; then, calculate dry seed weight (*see* **Note 15**).
7. Calculate the water contents (%) as follows:

Fig. 1 (**a**) Exemplar calendar using colored tape for each day of the week. The number on the label indicates the week number. The week number is determined by the first flowering day. (**b**) Tagged wheat spikes with seven-colored vinyl tape in the field

$$100 \times (\text{Fresh seed weight} - \text{Dry seed weight})/\text{Fresh seed weight}$$

3.3 Evaluation of Seed Dormancy Depth

1. Collect seeds as in Subheading 3.1, **step 4** (*see* **Note 16** and Fig. 4).
2. Place two sheets of filter paper inside a Petri dish (*see* **Note 17**).
3. Put approximately 20 seeds on the dish (*see* **Note 18** and Fig. 5).

Fig. 2 Representative photograph of flowering wheat. Anthers stick out from the middle part of the spike

Fig. 3 A wheat spikelet. First and second florets, located on the outermost spikelet, should be used for dormancy tests

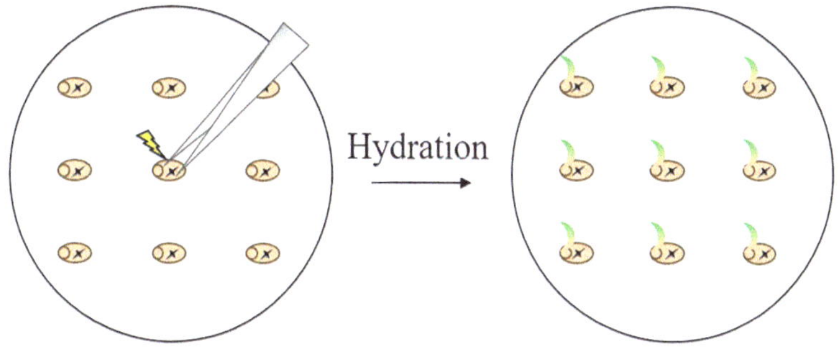

Fig. 4 An example of ineffective seed handling. Seeds wounded by sharp forceps germinate easily

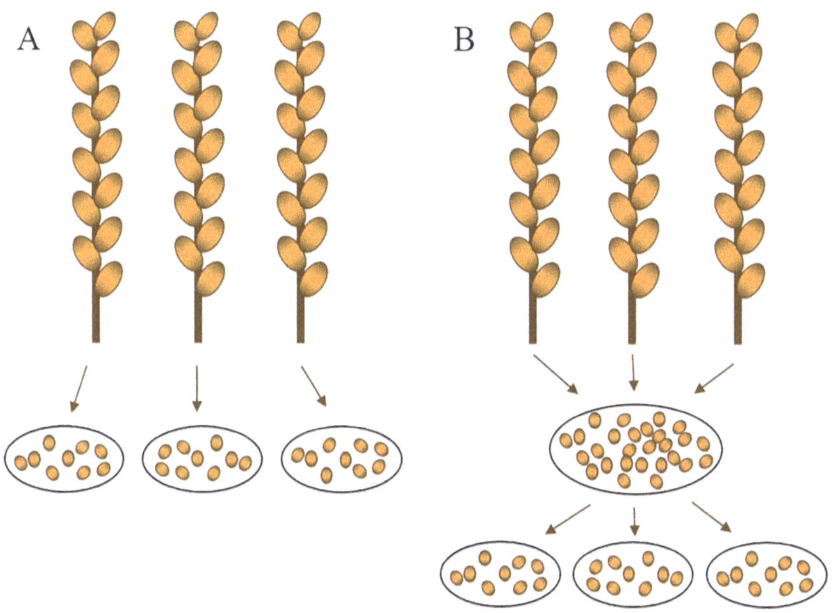

Fig. 5 Dormancy evaluation in seeds should be performed on spikes (**a**). Variability would be masked if seeds from several spikes are pooled (**b**)

4. Optional: Cut seeds into halves with a razor blade, and put approximately 20 halved seeds with the embryo on the dish (*see* **Note 4** and Fig. 6).
5. Pour 6 mL deionized water into the dish (*see* **Note 19**).
6. Optional: For ABA sensitivity test, pour 6 mL of a 10 μM ABA solution instead of deionized water (*see* **Note 3**).
7. Three replicates should be performed for statistical analysis.
8. Place dishes with seeds inside a plastic box with a lid (*see* **Note 20** and Fig. 7).
9. Incubate the box at 15 °C or 20 °C in the dark (*see* **Note 21**).

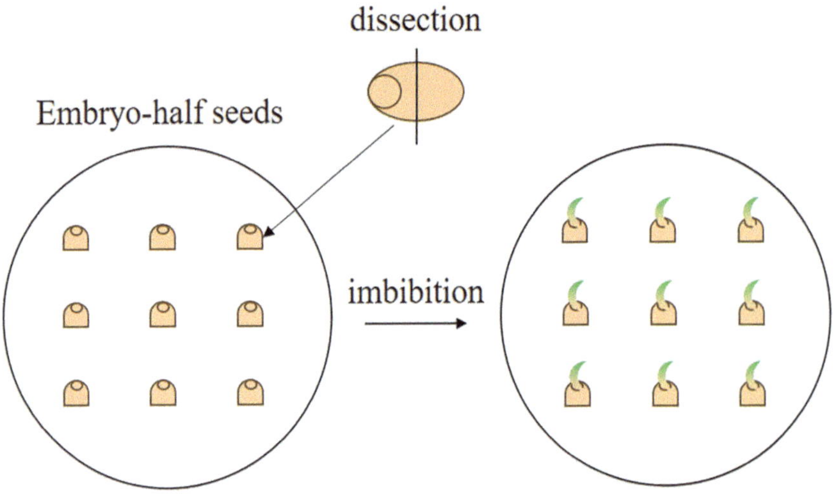

Fig. 6 Embryo-half seed germination test. Seeds are dissected, embryo-half seeds are hydrated, and then germination is checked

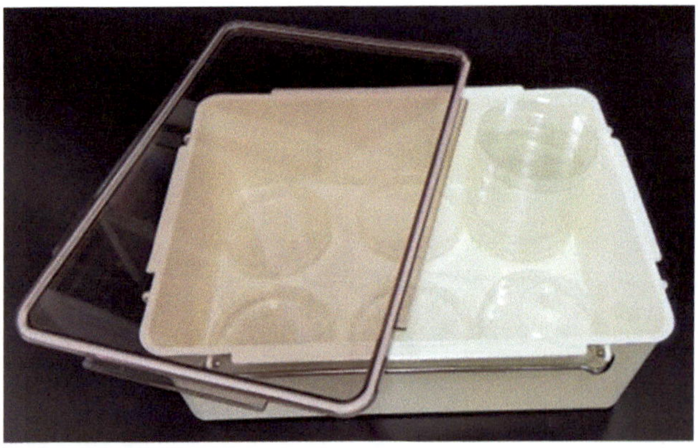

Fig 7 A plastic container (inner dimensions: 299 × 210 × 95 mm) for the germination test can accommodate six stacks of six Petri dishes (9-cm diameter); 36 dishes can be stacked in this box

10. Score germination of seeds every 24 h for 7 days. A seed is considered germinated when the shoot and radicle have emerged (Fig. 8).
11. Discard germinated seeds after checking.
12. Express germination as a cumulative value over time.
13. Optional: Calculate the germination index (GI) as follows:

$$GI = (7 \times N1 + 6 \times N2 + 5 \times N3 + 4 \times N4 + 3 \times N5 + 2 \times N6 + 1 \times N7) / (\text{Total days of the test} \times \text{Total seeds})$$

where N1 to N7 are the number of grains that germinated on days 1 to 7 (*see* **Note 22** and Fig. 9).

Fig. 8 Germinated seeds. Shoots and primary roots (radicles) emerge

3.4 Seed Color Distinction

1. Put several fully ripened seeds in a transparent test tube or a well of a 24-well plate.
2. Pour 1 mL of 1 N NaOH onto seeds (*see* **Note 23**).
3. Mix gently to submerge seeds in the solution at room temperature or 20 °C.
4. Observe seed color after 2 h (*see* **Note 24** and Fig. 10).

3.5 Tamyb10 *Genotyping*

1. PCR (20-μL volume): Combine 2 μL of 10 × *Ex Taq* Buffer (Mg^{2+} plus), 4 μL of 50% sterilized glycerol, 200 μM of each dNTP, 0.2 μM of each primer, DNA template (100 ng of genomic DNA), and 0.5 U TaKaRa Ex Taq HS.
   (a) Multiplex PCR 1 for *Tamyb10-A1* genotyping: Add primers for Tamyb10-LP6, Tamyb10-RP4, Tamyb10-LP7, Tamyb10-LP8, and ARP1. Denature at 94 °C for 2 min and then perform 30 cycles of amplification (30 cycles of 94 °C for 30 s, 61 °C for 30 s, and 72 °C for 1 min) and a final extension at 72 °C for 7 min.
   (b) Multiplex PCR 2 for *Tamyb10-B1* and *D1* genotyping: Add primers for Tamyb10-LP2, Tamyb10-RP1, DLP7, and DRP7. Denature at 94 °C for 2 min and then perform 30 cycles of amplification (30 cycles of 94 °C for 30 s, 63 °C for 30 s, and 72 °C for 20 s) and a final extension at 72 °C for 7 min.
2. Separate 12–20 μL of PCR products on a 1.5–2% (w/v) agarose gel for multiplex PCR 1 and 3% agarose gel for multiplex PCR 2.

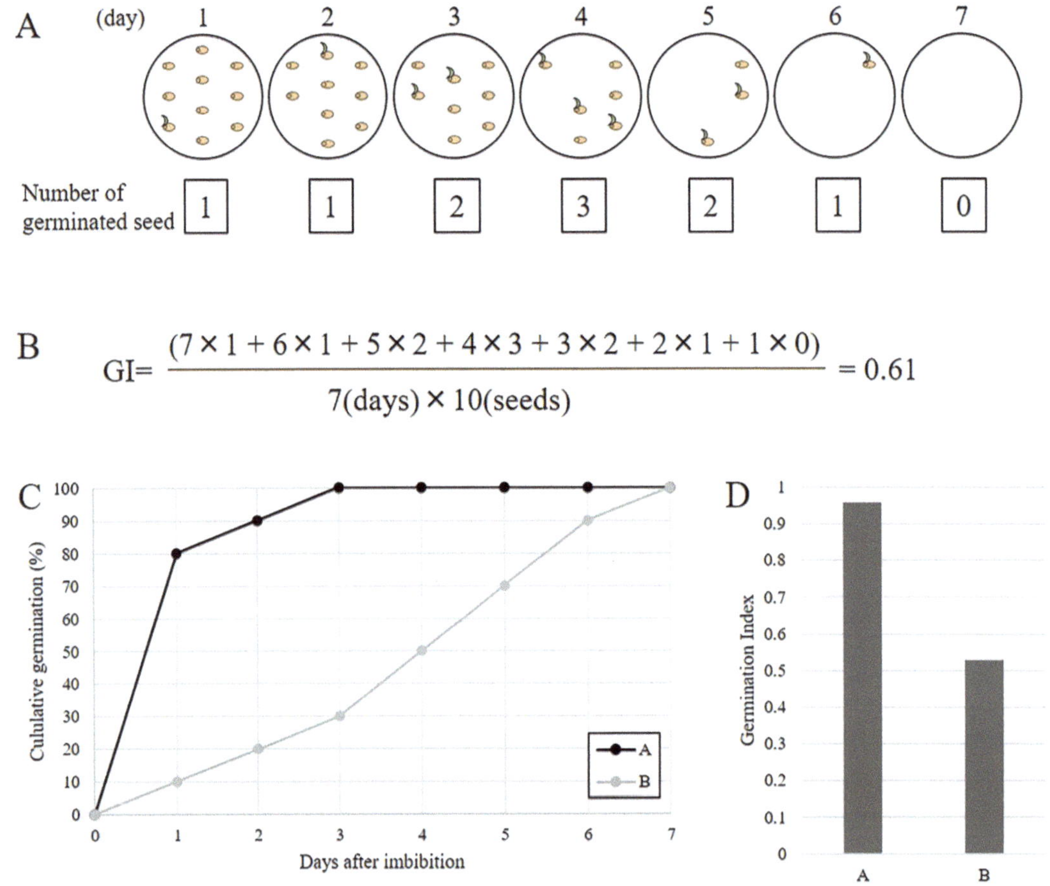

Fig. 9 Calculation of germination index (GI). Germination test example with 10 seeds for 7 days (**a**). If 1, 1, 2, 3, 2, 1, and 0 seed germinated on days 1 to 7, respectively, the corresponding GI will be 0.61 according to the formula shown in (**b**). Although two lines may have a germination rate of 100% (**c**), the GI can discriminate between them (**d**)

(a) Multiplex PCR 1 generates three amplification patterns, as illustrated in Fig. 11a:

(Type 1) a single fragment of 1060 bp corresponding to *R-A1b* (dominant),

(Type 2) a single fragment of 536 bp corresponding to *R-A1a* (recessive, CS-type), and (Type 3) two fragments of 1060 and 412 bp corresponding to *R-A1a* (recessive, Norin 17-type).

(b) Multiplex PCR 2 generates four amplification patterns, as illustrated in Fig. 11b:

(Type 1) two fragments of 282 and 209 bp corresponding to *R-B1b* (dominant)/*R-D1b* (dominant), (Type 2) a single fragment of 282 bp corresponding to *R-B1b* (dominant)/*R-D1a* (recessive), (Type 3) two

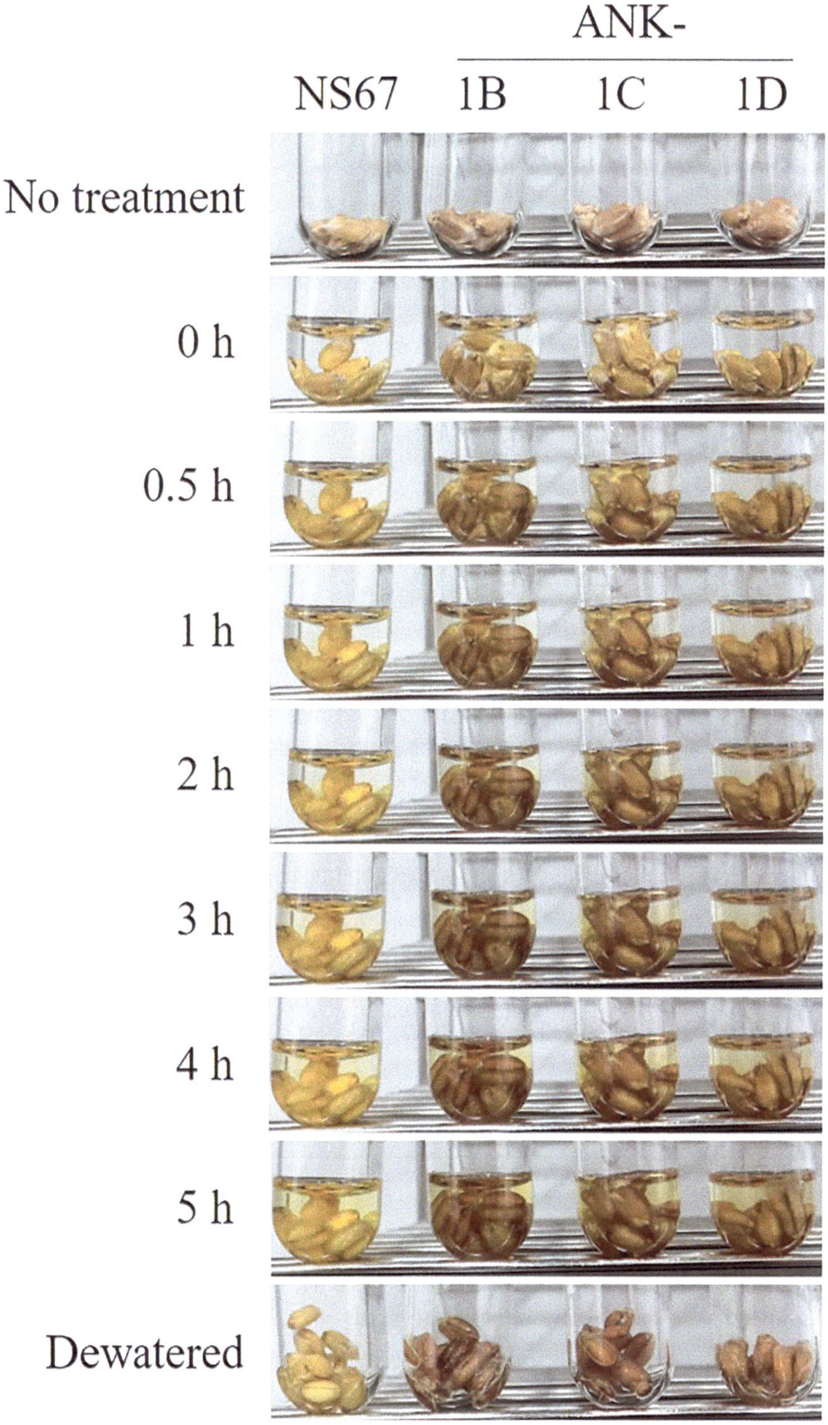

Fig. 10 Seed color detection of a white-grained line (Novosibirskaya 67; NS67) and its near-isogenic lines harboring *R-B1b* (ANK-1B), *R-D1b* (ANK-1C), and *R-A1b* (ANK-1D) [11]. Dry seeds (top). The addition of NaOH solution treatment to seeds from 0 h (after pouring) until 5 h. Dewatered seeds after 5 h of treatment (bottom)

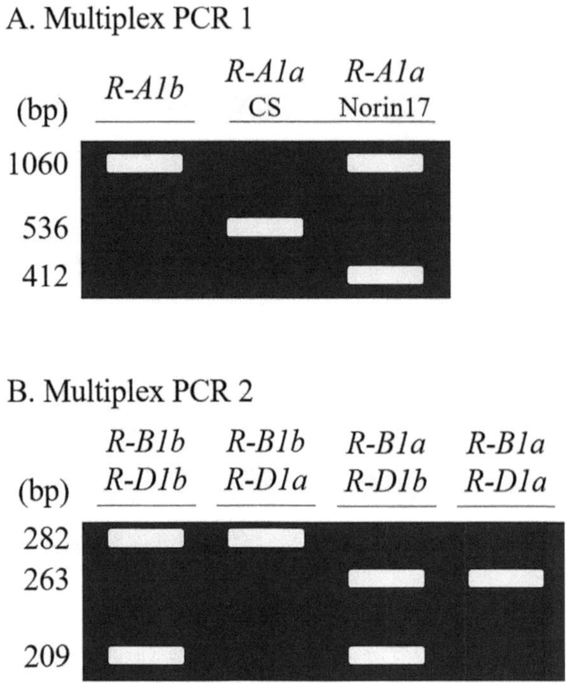

Fig. 11 (**a**) Diagram of 2% agarose gel electrophoresis for multiplex PCR 1. *R-A1b* (dominant) lines show a single fragment of 1060 bp, whereas *R-A1a* (recessive) lines have a single fragment of 536 bp or two fragments of 1060 and 412 bp. (**b**) Diagram of 3% agarose gel electrophoresis for multiplex PCR 2. *R-B1b* (dominant) lines show a 282-bp fragment, whereas *R-B1a* (recessive) lines show a 263-bp fragment. Only *R-D1b* (dominant) lines show a 209-bp fragment

fragments of 263 and 209 bp corresponding to *R-B1a* (recessive)/*R-D1b* (dominant), and (Type 4) a single fragment of 263 bp corresponding to *R-B1a* (recessive)/*R-D1a* (recessive).

4 Notes

1. Each of the seven colors of vinyl tape corresponds to a different day of the week. Tape with vivid colors is preferable to recognize colors in the field.
2. Filter paper (8.4-cm diameter) can be ordered specially, for example, ADVANTEC No. 2 84-mm from Toyo Roshi Kaisha, Ltd. Japan.
3. The sensitivity of seeds to ABA should be tested using embryo-half seeds, as cutting the seed in half will release dormancy.

4. A seed that does not germinate can be either dead or dormant. However, using embryo-half seeds for germination tests can distinguish between the two possibilities, as the inflicted wound will break dormancy. Embryo-half seeds germinate if the seed was dormant. Therefore, germination should be checked for both whole and embryo-half seeds (Fig. 6).
5. NaOH should be weighed rapidly because NaOH is a deliquescent solid.
6. A stock solution of NaOH should be stored in plastic containers because glass reacts slowly with aqueous NaOH solutions to form soluble silicates.
7. Tween 20 helps the NaOH solution penetrate the seeds.
8. For successful PCR, using a DNA extraction kit, e.g., DNeasy Plant Mini Kit (Qiagen), is highly recommended.
9. *Tamyb10* genes are difficult to amplify using regular Taq polymerases, e.g., Ex Taq, since they have high GC contents. In this protocol, glycerol was added to the solution to solve this problem, but it is unnecessary if polymerases optimized specifically for difficult templates are used.
10. If the wheat line starts flowering on a Sunday, tag with red tape and write "1" on it, as illustrated in Fig. 1a.
11. Add a piece of vinyl tape to tag spike necks, as show in Fig. 1b.
12. Florets flower gradually to the end of the spike (Fig. 2). Flowering time will thus differ between the central and distal florets, even in the same spike. Therefore, only seeds from the central spikelets of a spike should be used. In addition, after the third florets, smaller seeds develop outside of the first and second florets but should not be used. Only use seeds in the first and second florets (Fig. 3).
13. For accuracy, weigh seeds in batches of 11 or more, rather than just a few seeds.
14. Baked grains absorb moisture from the air. Therefore, weigh the grains immediately after cooldown to room temperature or keep in a desiccator until weighing.
15. Weighing empty dishes before and after baking can increase accuracy of results. Therefore, check if the dishes weigh the same before and after baking.
16. Do not freeze developing seeds for germination tests, as seeds will die. Freshly harvested seeds should be used immediately for germination tests. Also, seeds should be handled with care and without sharp forceps to prevent wounding, which would break dormancy independently of seed status. If the results of the germination tests are variable, check your forceps (Fig. 4).

17. Use new Petri dishes to avoid the effect of α-amylase secreted by seeds from the previous experiment.
18. The number of seeds to be used is not strict. Therefore, do not mix seeds from different spikes to make up for seed number. Seed dormancy should be evaluated for each spike, as illustrated in Fig. 5.
19. Too much water inhibits germination because of water sensitivity. The volume of 6 mL is adjusted for 20 seeds placed on a 9-cm-diameter Petri dish with two sheets of filter paper.
20. A box with inside dimensions of 299 × 210 × 95 mm can store 36 dishes (Fig. 7).
21. A strict incubation temperature is required because temperature influences the release of seed dormancy. Lower temperatures easily break dormancy. Low-dormancy lines will germinate rapidly at 20 °C, whereas high-dormancy lines will not germinate at 15 °C even 7 days after seed hydration. Although incubation is performed in the dark, checking the germination out of darkness within a few minutes is feasible.
22. The GI gives greater importance to seeds that germinate rapidly. If all seeds of two lines germinate 100% after 7 days, the GI will reflect differences in germination (Fig. 9).
23. If seeds do not thoroughly soak, add more NaOH solution.
24. After soaking in NaOH for 30 min, seed color should be observed (Fig. 10). However, do not soak seeds overnight because seeds will collapse.

References

1. Chmelar F, Mostovoj K (1938) On the application of some old and the introduction of new methods for testing genuineness of variety in the laboratory. Proc Int Seed Test Assoc 10: 67–74
2. DePauw RM, McCaig TN (1988) Utilization of sodium hydroxide to assess kernel color and its inheritance in eleven spring wheat varieties. Can J Plant Sci 68(2):323–329. https://doi.org/10.4141/cjps88-042
3. Biffen RH (1905) Mendel's laws of inheritance and wheat breeding. J Agric Sci 1(1):4–48. https://doi.org/10.1017/S0021859600000137
4. McIntosh RA, Yamazaki Y, Dubcovsky J, Rogers J, Morris C, Somers DJ et al (2010) National BioResource Project (NBRP)/KOMUGI. http://www.shigen.nig.ac.jp/wheat/komugi/genes/download.jsp
5. Himi E, Maekawa M, Miura H, Noda K (2011) Development of PCR markers for *Tamyb10* related to *R-1*, red grain color gene in wheat. Theor Appl Genet 122(8):1561–1576
6. Himi E, Nisar A, Noda K (2005) Colour genes (R and Rc) for grain and coleoptile upregulate flavonoid biosynthesis genes in wheat. Genome 48(4):747–754
7. Kurihara-Yonemoto S, Matsunaka H (2019) A simple method for genotyping of wheat seed coat color genes by multiplex PCRs. Bull NARO Agric Res Hokkaido Reg 207:35–49. [In Japanese with English summary, tables, and figures]. https://doi.org/10.24514/00002276
8. Nilsson-Ehle H (1914) Zur Kenntnis Der Mit Der Keimungsphysiologie Des Weizens in Zusammenhang Stehenden Inneren Faktoren. Zeitschrift fur Pflanzenzuchtung 2:153–187

9. Flintham JE (2000) Different genetic components control coat-imposed and embryo-imposed dormancy in wheat. Seed Sci Res 10: 43–50
10. Walker-Simmons M (1987) ABA levels and sensitivity in developing wheat embryos of sprouting resistant and susceptible cultivars. Plant Physiol 84:61–66
11. Koval SF (1997) The catalog of near-isogenic lines of Novosibirskaya-67 common wheat and principles of their use in experiments. Russ J Genet 33(8):995–1000

Chapter 5

Assessing Seed Germination Response of Parasitic Plant *Striga hermonthica* with Small-Molecule Probes

Jia Xin Yap and Yuichiro Tsuchiya

Abstract

Seed germination of a parasitic plant *Striga hermonthica* is elicited by strigolactones which are exuded from roots of host plants. Here, we describe a high-throughput germination assay and a method for visualizing in vivo strigolactone receptor functions with a fluorogenic probe.

Key words Parasitic plant, *Striga hermonthica*, Suicidal germination, Chemical screening, Yoshimulactone green W, Fluorescent molecular probe, Strigolactone, Receptor function visualization, Time-lapse movie

1 Introduction

Parasitic plant *Striga hermonthica* (*Striga*) is a noxious weed that causes massive damages to crop production in Africa through parasitizing important crops including sorghum and maize [1]. *Striga* has adapted a unique parasitic life cycle, such as its seed germination is dependent on strigolactones (SLs) exuded from the roots of host plants [2]. Since seed germination is a key step for successful parasitism of *Striga*, stable methods for assaying this process would help to provide useful insights to accelerate the development of solutions for parasitic plant control. A current promising approach to combat *Striga* infestations utilizes synthetic SLs to force *Striga* seeds to germinate in the absence of host plants and eventually lead them to death due to nutrient deprivation. This strategy called "suicidal germination" requires highly potent synthetic SLs. During the initial phase of developing such suicide germination inducers, performance of the candidate molecules is often evaluated by high-throughput in vitro germination assay in 96-well plates.

Despite the tiny seed size of 0.3 mm in diameter, *Striga* stay alive for over 20 years in dormant state as dry seeds even in the soil.

Naoto Kawakami and Kazuhiro Sato (eds.), *Seed Dormancy: Methods and Protocols*, Methods in Molecular Biology, vol. 2830, https://doi.org/10.1007/978-1-0716-3965-8_5,

While the dry seeds are unable to respond to SLs immediately after imbibition, a period of seed conditioning under warm moist treatment gradually increases the sensitivity of *Striga* seeds toward the host-derived germination cues through upregulation of *HYPO-SENSITIVE TO LIGHT/KARRIKIN INSENSITIVE2 (HTL/KAI2)* genes that encode a family of SL receptors [3]. Recent small-molecule technology enables the visualization of the function of ShHTLs in germinating *Striga* seeds. The small-molecule probe, known as Yoshimulactone green W (YLGW), is a fluorogenic substrate of ShHTLs, which upon hydrolysis releases fluorescent moiety [3]. This attribute enables to visualize live images of SL perception in germinating seeds under fluorescent microscopes, allowing detailed understanding of germination processes and how small-molecule regulators affect them.

Here, we share protocols for high-throughput germination assay applicable for screening thousands of small-molecule from chemical libraries, pot germination assay, and a method for visualizing SL perception with YLGW under fluorescent microscopes.

2 Materials

2.1 Striga *Seed Conditioning*

1. *Striga* seeds.
2. Aluminum foil.
3. Kitchen bleach or sodium hypochlorite.
4. Ultrasonic bath sonicator.
5. 35 mm culture dish.
6. Glass fiber filter paper (Whatman GF/B 2.5 cm).
7. Parafilm.
8. Clean bench.
9. Incubator.

2.2 *In Vitro* Striga *Germination Assay*

1. Conditioned *Striga* seeds.
2. 96-well plate.
3. Small-molecule of interest.
4. DMSO or appropriate solvent for dissolving small-molecules.
5. Microscope.

2.3 Striga *Germination Assay in Pots*

1. Dry *Striga* seeds.
2. Nylon mesh with 0.1 mm sieve.
3. Pots.
4. Heat-sealer.

2.4 YLGW Time-Lapse Imaging

1. Conditioned *Striga* seeds.
2. Sterilized 1% agarose.
3. 35 mm glass bottom dish (Matsunami, D11130H, glass diameter 14 mm, uncoated).
4. Sterilized toothpick.
5. Sterilized filter paper.
6. Tissue.
7. Inverted laser microscope (Leica DMi8).
8. YLGW.

3 Methods

Striga seeds should be handled carefully in all steps to prevent spreading. The use of *Striga* seeds should be recorded appropriately according to the rules and regulations of each country.

3.1 Striga *Seed Conditioning*

1. Weigh 0.05 g of *Striga* dry seeds in a 1.5 mL micro tube (*see* **Note 1** and Fig. 1).

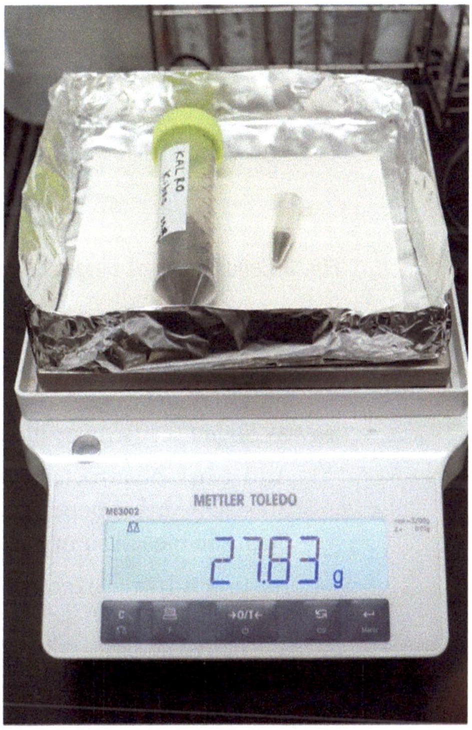

Fig. 1 Weighing *Striga* seeds. To prevent the spreading of seeds, weighing is carefully carried out on an aluminum foil with four sides folded up to form a tray to prevent the spreading of seeds

Fig. 2 Surface sterilization of *Striga* seeds. Compared with nonsterilized seeds (left), the color of seeds become light brown after sterilization (right)

Fig. 3 Conditioning of *Striga* seeds on glass fiber filter paper in 35 mm culture dish

2. Add 1 mL of 20% bleach solution and vortex for 1 min. Discard supernatant. Repeat the step four more times or until the color of *Striga* seeds turns light brown (*see* **Note 2** and Fig. 2).
3. Add 1 mL of autoclaved Milli Q and sonicate in water bath for 30 s. Discard supernatant. Repeat the step nine more times or until the bleach solution is completely washed out (*see* **Note 3**).
4. Resuspend sterilized *Striga* seeds in 1 mL of autoclaved Milli Q.
5. At the clean bench, place one glass fiber filter paper in a 35 mm culture dish (*see* **Note 4**).
6. Spread 0.6 mL of sterilized *Striga* seeds on the glass fiber filter paper (*see* **Note 5** and Fig. 3).

7. Seal the culture dish with two layers of parafilm and wrap with a layer of aluminum foil (*see* **Note 6**).
8. Incubate the dish at 30 °C in dark for1–2 weeks before use (*see* **Note 7**).

3.2 *In Vitro* Striga *Germination Assay*

1. Add 1 μL per well of 100× stock solution of small-molecule to a 96-well plate. Use the same amount of solvent control or use a final concentration of 0.1 μM rac-GR24 (racemic mixture of synthetic strigolactone, GR24) as the positive control (*see* **Note 8**).
2. Scoop the conditioned *Striga* seeds from the filter paper using a spatula and resuspend in 1 mL of autoclaved Milli Q in a 1.5 mL Eppendorf tube. Add 99 μL of approximately 20 to 30 *Striga* seeds to a final volume of 100 μL per well.
3. Shake the plate to mix well and seal the plate with two layers of parafilm.
4. Incubate the plate at 30 °C in dark for 2 days (*see* **Note 9** and Fig. 4).
5. Observe and count the germinated *Striga* seeds (seeds with emerged radicle) of each well under a microscope.
6. Calculate the average germination rate from three replicates in Microsoft Excel (*see* **Note 10**).

3.3 Striga *Germination Assay in Pots*

1. Prepare 3 cm × 3 cm square envelope with nylon mesh with three edges sealed by a heat-sealer.
2. Pore around 100 seeds in the envelope and seal the opened edge.
3. Fill pots with soil and bury the envelope, add water, and cover the top with thin foil.

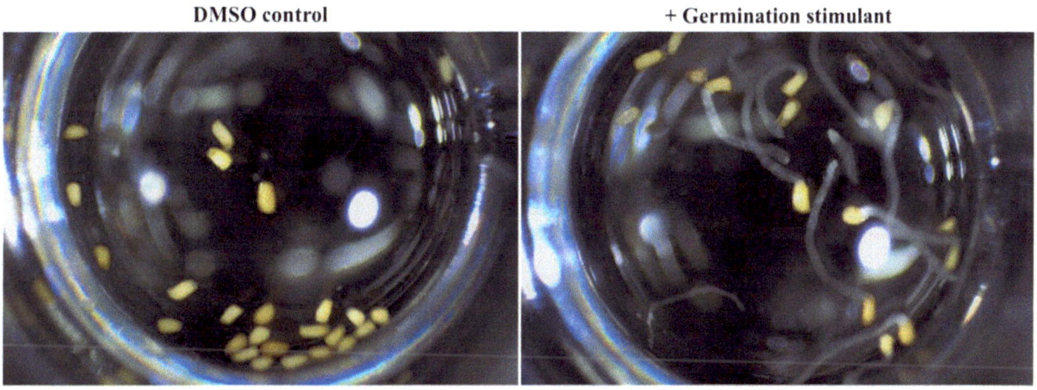

Fig. 4 *Striga* seeds treated with DMSO control or germination stimulant such as 0.1 μM GR24 in a 96-well plate. Images were taken on 2 days after small-molecule application

4. Condition seeds at 30 °C for 1 week with moist condition.
5. Apply small-molecule solution onto the soil and let seeds germinate for 2 days.
6. Take the envelope out from the soil and rinse with water.
7. Move the seeds to petri dish filled with water.
8. Count germination under dissecting microscope.

3.4 YLGW Time-Lapse Imaging

1. Add 200 μL per well of autoclaved 1% agarose solution to a 48-well plate and cool down until completely solidified (*see* **Note 11**).
2. Transfer the agarose gel block onto a glass bottom dish using a spatula (*see* **Note 12** and Fig. 5).
3. Resuspend conditioned *Striga* seeds in 1 mL of autoclaved Milli Q in a 1.5 mL Eppendorf tube.
4. Place a filter paper on the lid of a 35 mm culture dish and add 200 μL of *Striga* seeds onto the filter paper (*see* **Note 13**).
5. Using a toothpick, gently press a seed from the end of its cotyledon to remove the seed coat and obtain the seed embryo under a microscope. Place the seed embryo onto the agarose gel block using the toothpick. Repeat the step until 30 embryos are obtained (*see* **Note 14** and Fig. 6).
6. Invert the agarose gel block containing seed embryos upside down using a spatula. Wet a tissue with Milli Q and place it at the edge of the glass bottom dish surrounding the agarose gel block (*see* **Note 15** and Fig. 7).
7. Set up the condition for taking germination movie with bright field and GFP lenses of a laser microscope at 20 min interval for 72 h (*see* **Note 16**).

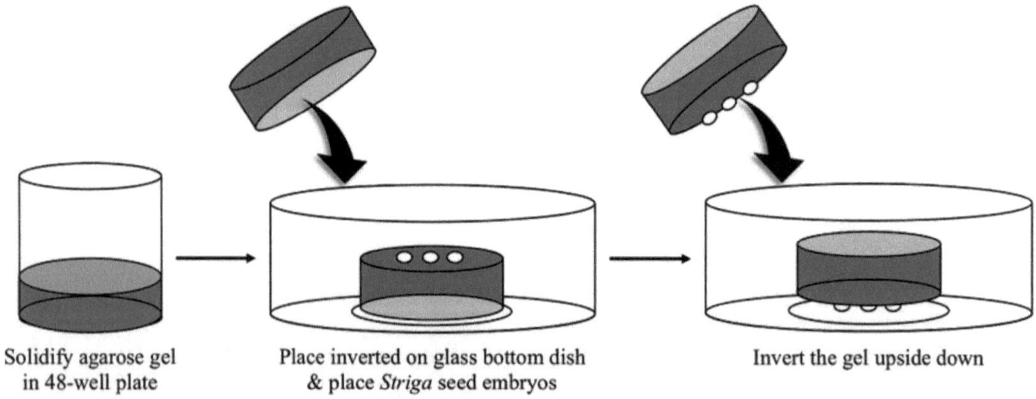

Fig. 5 Preparation of *Striga* seed embryos on gel block for taking time-lapse movie on glass bottom dish

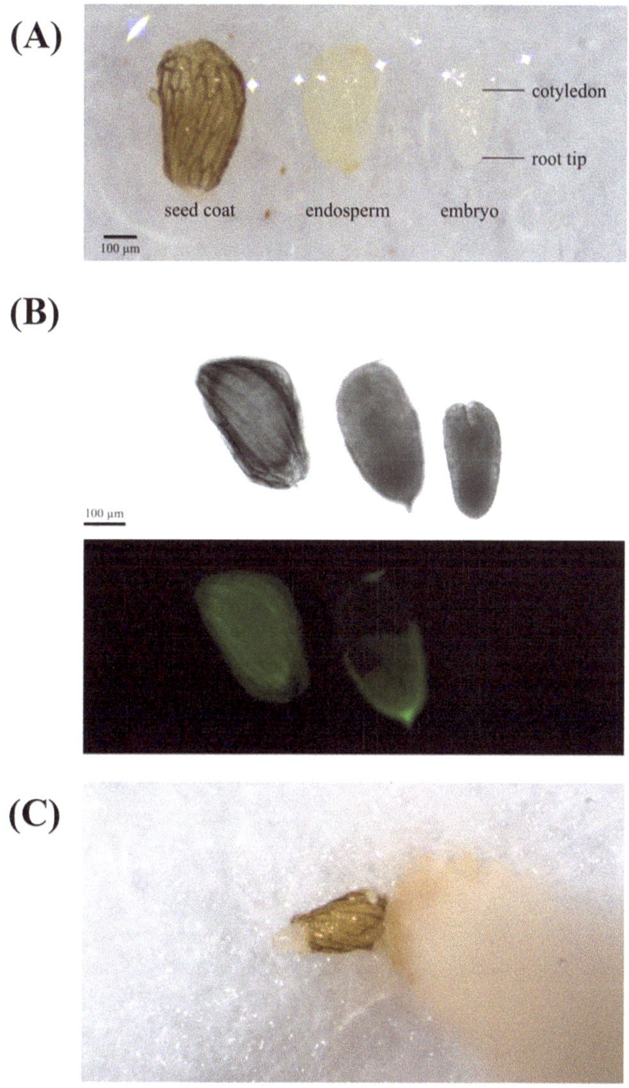

Fig. 6 (**a**) Dissection of embryo from seeds. The embryo is encapsulated in a layer of endosperm that is covered by the seed coat. (**b**) Auto-fluorescence of intact *Striga* seed (left), endosperm layer (middle), and embryo (right). Image was taken with FITC filter. (**c**) Embryo can be pushed out from the endosperm by gently applying force from the cotyledon side using a toothpick

8. Prepare 10 µM YLGW solution by dissolving 1 µL of 10 mM YLGW in 1 mL of autoclaved Milli Q. Add 20 µL of 10 µM YLGW solution from the edge of the agarose gel block, cover the lid of the glass bottom dish, and start the movie (*see* **Note 17**).
9. Export the germination movie and analyze YLGW fluorescence wave using kymograph function of ImageJ (*see* **Note 18** and Figs. 8 and 9).

Fig. 7 Sample setup for YLGW time-lapse imaging. Gel block with dissected *Striga* embryos is placed in the center of glass bottom dish and surrounded by a layer of wetted tissue to prevent the gel from drying out throughout imaging

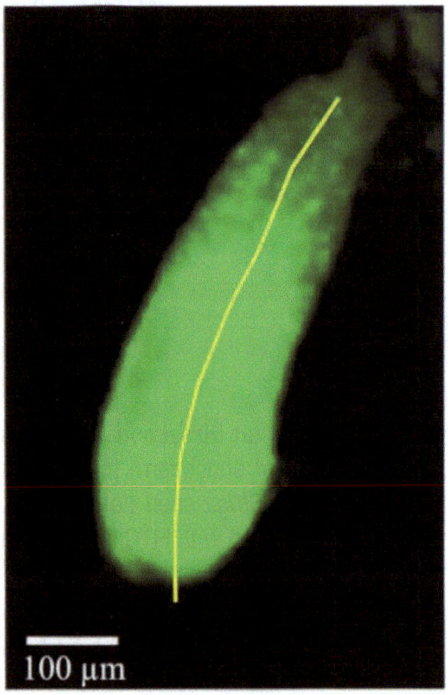

Fig. 8 Kymograph analysis of YLGW time-lapse imaging. The center from cotyledon to root is plotted with segmented line function of ImageJ on the last stack of time-lapse images, and the stacks are converted into kymograph for subsequent analysis

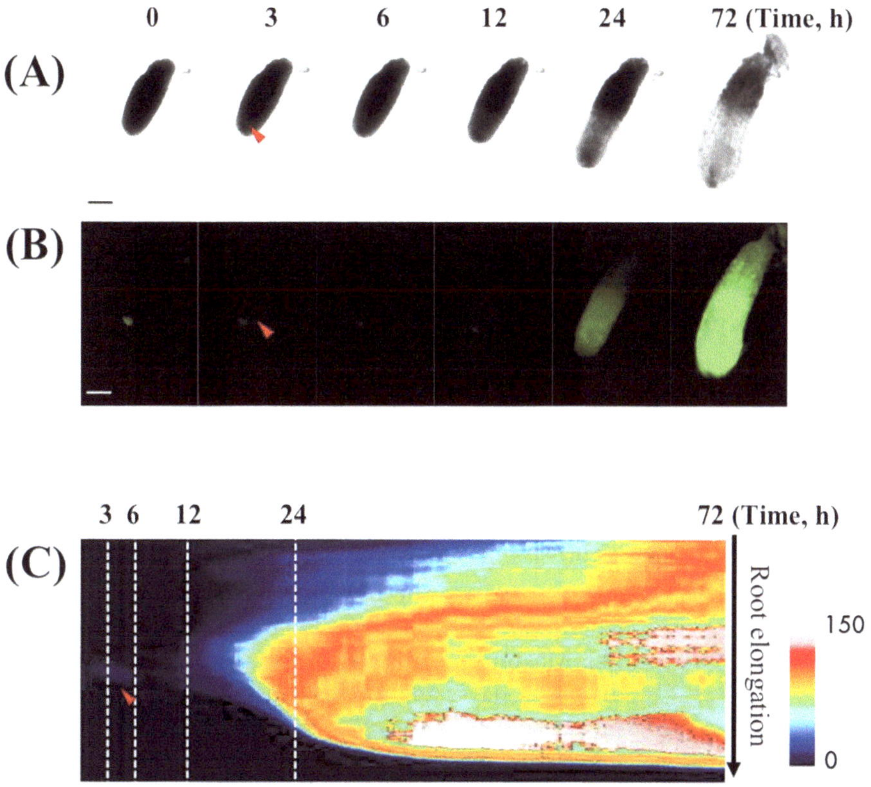

Fig. 9 Dynamics of SL perception during germination indicated by fluorescence from YLGW hydrolysis by ShHTLs taken with (**a**) brightfield and (**b**) FITC filter of Leica DMi8. (**c**) Kymograph analysis converted from time lapse movie of *Striga* seed germination. Red arrowhead represents appearance of first wake-up wave. Bar represents 100 μm

4 Notes

1. We usually keep the *Striga* seed stocks in 50 mL falcon tubes and tap the desired amount of use carefully into Eppendorf tube on an aluminum foil with four sides folded up to form a tray to prevent the spreading of seeds (Fig. 1). We recommend using a heavy-duty foil or two layers of regular foil, and add a layer of tissue on top for easier visualization of seeds that drop out during the transfer. *Striga* dry seeds (0.05 g) are usually sufficient for germination assay in two 96-well plates.
2. We use commercially available kitchen bleach for seed surface sterilization and dilute freshly before use, but it can be substituted with appropriate amount of sodium hypochlorite. *Striga* dry seeds collected from the field are usually covered with soil and dark in color. Insufficient wash often causes fungal contamination during conditioning period whereas over wash

kills the seeds, which in either way will affect the results of germination assay. We usually stop bleaching by judging at the change of color from dark to light brown as shown in Fig. 2.

3. We highlight the importance of complete removal of bleach as it will hydrolyze SLs and possibly other small-molecules when stay on the seed coat. We find washing by water bath sonication gives the best result, but this step can be replaced by vortex and flicking the tube with hand until all the air bubbles formed from the bleach solution completely disappear.
4. We recommend this step to be done at clean bench to prevent contamination.
5. We highlight the importance of optimum moisture during conditioning for obtaining high *Striga* germination rate.
6. Sealing with parafilm helps to prevent seeds from drying out throughout the conditioning period. An example of plate setup is shown Fig. 3.
7. *Striga* seeds are the most sensitive to germination stimulants when conditioned at 30 °C compared to lower temperatures [4]. Maximum germination rate usually peaks after 7–21 conditioning days, and prolonged conditioning period (28 days onward) likely induces secondary dormancy and reduces germination. Besides, conditioning should be done in dark because exposure to light exerts inhibition on *Striga* germination.
8. We use DMSO as the solvent to dissolve small-molecules, and final concentration of DMSO or any other solvent used should be below 1% to not affect biological activities. Other reported SLs or SL analogs can be used as positive control instead of rac-GR24.
9. Radicle protrusions usually occur after 24 h, but we recommend 2-day incubation for scoring the maximum germination and clearer observation. Figure 4 shows representative images of *Striga* seeds treated with DMSO control or germination stimulant in a 96-well plate on day 2.
10. *Striga* germination is solely dependent on germination stimulants like SLs, and maximum germination rate of conditioned seeds is usually around 50%–90% under laboratory conditions.
11. We find gel block prepared in 48-well plate a good size for glass bottom dish with glass diameter of 14 mm and sufficient to place up to 30 *Striga* seed embryos. Otherwise, it can be prepared by cutting a gel into approximately 5 mm squares. Agarose solution (1%) can be prepared beforehand, stored up to a few months at room temperature, and dissolved in microwave before use. The size of gel block prepared in a 48-well plate fits well for glass bottom dish with a glass diameter of

14 mm. This gel size is sufficient to hold up to 30 *Striga* seed embryos. Alternatively, gel block can be prepared by cutting a gel into approximately 5 mm squares. Agarose solution (1%) can be prepared beforehand, stored up to a few months at room temperature, and dissolved in microwave before use.

12. The solidified agarose gel block can be taken out from the 48-well plate using a spatula and placed on the glass bottom dish in an upside down manner to put the seed embryos on the flat surface of the gel block. After putting the embryos onto the gel, invert upside down again to place the embryos onto the glass bottom as summarized in Fig. 5.
13. We use here the filter paper for Western blotting.
14. The size and shape of *Striga* seeds usually vary greatly, and we recommend selecting seeds with bigger embryo. The seeds are generally irregular ovate shape, where the wider end is usually cotyledon and narrower end is the root tip. The embryo is encapsulated in a layer of endosperm that is covered by the seed coat (Fig. 6a).

    The seed coat and endosperm emit autofluorescence when excited at the same wavelength for YLGW and therefore have to be removed before taking time-lapse movie (Fig. 6b).

    The embryo can be pushed out from the endosperm by gently applying force from the cotyledon side using a sharp tip like a toothpick or a dissecting needle (Fig. 6c). Otherwise, the seeds can be first cut with a razor at either end and then push out the embryos with toothpick or needle. However, be careful not to cut off any part of the embryo especially the root tip that is important for the initial signal perception during germination [3].

    After removing its seed coat and endosperm, the embryo can be picked up easily with the same toothpick or needle to transfer from filter paper onto agarose gel block.
15. To prevent the agarose gel from shrinking throughout the period of time for taking time-lapse movie, a layer of wetted tissue is added around the gel block to provide moisture as shown in Fig. 7.
16. For Leica DMi8, we use 4× objective lens and tilling function in LasX software to cover the whole area of 30 embryos. Be sure to estimate enough extra area for germinating embryos. For brightfield, exposure time is set at 10 ms. For tracking fluorescence emission from YLGW, we use FITC filter with exposure time of 1 s (or longer exposure of 1.5 s is recommended for observing initial wake-up wave). We also set multiple Z-stacks and select the best focus for movie analysis.
17. Be sure to remove the lid of 35 mm glass bottom dish during the microscope setup for the ease of YLGW addition. After

adding YLGW solution, cover with the lid to prevent the agarose gel from drying out. Try not to move the dish during this step to ensure the area set for taking time-lapse movie is not affected.

18. Open the YLGW fluorescence time-lapse movie in ImageJ and convert into grayscale. Scroll to the last stack and select an embryo to draw a line at the center from cotyledon to root using segmented line function (yellow line in Fig. 8). Select analyze > multi kymograph > multi kymograph to convert time-lapse movie into a kymograph image. We usually apply LUT > royal to represent the YLGW fluorescence.

References

1. Ejeta G (2007) The Striga scourge in Africa: a growing pandemic. In: Integrating new Technologies for Striga Control: towards ending the witch-hunt. World Scientific, Hackensack, pp 3–15
2. Xie X, Yoneyama K, Yoneyama K (2010) The strigolactone story. Annu Rev Phytopathol 48: 93–117
3. Tsuchiya Y, Yoshimura M, Sato Y, Kuwata K, Toh S, Holbrook-Smith D et al (2015) Probing strigolactone receptors in Striga hermonthica with fluorescence. Science 349:864–868
4. Aflakpui GKS, Gregory PJ, Froud-Williams RJ (1998) Effect of temperature on seed germination rate of Striga hermonthica (Del.) Benth. Crop Prot 17(2):129–133

Chapter 6

A Seed Storage Protocol to Determine Longevity

Julia Buitink and Olivier Leprince

Abstract

The longevity of seeds, also known as storability, is the period of time for which a seed lot maintains its viability during storage. The method aims to determine longevity of a seed lot during storage in a controlled environment. Seeds are first rehydrated to a preset water content (or relative humidity, RH) and then incubated under controlled conditions for various periods of time to allow for deterioration to occur. At increasing intervals during storage, seeds are retrieved and viability is tested by scoring germination of the seed lot (i.e., radicle protrusion). From these data, a survival curve can be drawn depicting loss of germination during time of storage from which different parameters estimating longevity can be inferred. These parameters can be used to compare longevity between different seed lots, genotypes, or species at similar storage conditions. This test can also be used as a proxy to measure seed vigor or physiological seed quality.

Key words Aging, Deterioration, Equilibration, Germination, Longevity, Relative humidity, Seed, Storability, Temperature, Viability

1 Introduction

Longevity or storability of seeds is a measure of how long seeds remain viable during storage under a given set of controlled conditions. These conditions are set to promote deterioration within the cells and hence loss of viability. Deterioration is a time-dependent and stochastic phenomenon. Therefore, assessing longevity must encompass a time component rather than assessing viability at a single time point during storage [1]. The most important parameters that influence aging during storage are seed water content, storage temperature, and oxygen. Their choice is important as different combinations of these three factors will influence the rate of seed aging and hence the duration of the experiment but also the type of aging reactions [2, 3]. Seed water content is manipulated by exposing seeds to different relative humidities (RH) that are generated using salt solutions [4, 5]. Next to storage conditions, longevity depends on the genetic background [6–9]

Naoto Kawakami and Kazuhiro Sato (eds.), *Seed Dormancy: Methods and Protocols*, Methods in Molecular Biology, vol. 2830, https://doi.org/10.1007/978-1-0716-3965-8_6,

and prestorage factors such as level of seed maturity at harvest and seed conditioning after harvest treatment [9–11].

Correct assessment of longevity is important for gene bank curators in charge of storing seeds for long periods and monitoring viability to determine regeneration frequencies and for academics investigating the role of genes in longevity via functional analysis of mutants. Therefore, a consistent approach is needed to be able to compare results between experiments. Here, we propose a protocol based on the generation of seed survival curves. The different steps that are described are seed conditioning prior to storage, storing seeds in conditions that are deleterious enough for deterioration within a reasonable time scale (months instead of decades) while remaining appropriate to the potential downstream applications of the findings, determination of viability over time of storage, and data analysis to infer longevity.

2 Materials

2.1 Seed Conditioning

1. Hermetically sealable containers wherein air can freely circulate to dry or rehydrate seeds.
2. Relative Humidity (RH) meter.
3. Saturated salt solution for 44% RH: dissolve K_2CO_3 in water and keep adding salt until a slurry is formed. Place the slurry in a container and close hermetically. The volume of the slurry should occupy 10–20% of the volume of the container (*see* **Note 1**). Let equilibrate for 3 days at room temperature. Thereafter, a small layer of saturated water will form above the salt crystals. The slurry needs to be checked every 1–-2 months to ensure that no drying or hydration occurred (due to exchange with seed moisture). In this case, adjust by adding water or salt (*see* **Note 2**).
4. Salt solution for 60% RH: dissolve 300 g LiCl in 1 L water with a magnetic stirrer. Transfer the LiCl solution to a container and close hermetically (*see* **Note 3**). Prepare all salt solutions 3 days in advance of the storage experiment.

2.2 Storage Experiment

1. Temperature-controlled incubator or fan-assisted oven.
2. Moisture-proof packets such as laminated aluminum bags and sealer for closed storage.
3. Hermetically sealable containers for open storage.

2.3 Germination Tests

1. Whatman no. 1 filter paper.
2. Petri dishes, pleated paper, or germination rolls depending on seed size.

3. Temperature-controlled cabinets with or without lighting, depending on the germination requirement for the species studied.
4. Demineralized water.

3 Methods

3.1 Seed Conditioning

1. Count 5–10 samples of 50–100 seeds from at least 3 plants that represent different biological replicates, depending on seed availability (*see* **Notes 4–6**).
2. Place each sample in open glass Petri dish of suitable size (*see* **Note** 7). If seeds need scarification, chipping, or dehusking, perform this prior to the equilibration step to allow water exchange between the atmosphere and seed tissues.
3. Freshly harvested seeds first need to be dried to a lower RH than the storage experiment RH (*see* **Notes 8–9**). Petri dishes containing seeds are placed on a mesh above a saturated salt solution at 44% RH (K_2CO_3) at 20 °C–25 °C (*see* **Note 2**). Leave the system to equilibrate for 7 to 14 days depending on the seed size (*see* **Note 10**).
4. Transfer the open Petri dishes containing the seeds above the salt solution at 60% RH, close the lid, and incubate for 7 days (*see* **Note 10**) at room temperature (*see* **Notes 3** and **11**). Keep one sample to test for viability before starting the storage experiment to test whether the seed lot is dormant, desiccation sensitive, or contains immature seeds. This will also constitute the time point zero days of storage.

3.2 Survival Monitoring during Storage

1. If closed storage is considered, retrieve the samples from the equilibration box and place them in laminated aluminum bags, remove as much air as possible, and seal thereafter. Place the bags at 45 °C (*see* **Notes 12** and **13**). If open storage is considered, leave the seed samples in the equilibration box and place it at 45 °C.
2. At different time intervals (*see* **Note 14**), take one sample out of storage. Let the storage container/bag cool down to room temperature prior to opening.
3. Prepare one Petri dish per biological sample with one Whatman filter paper and add distilled, sterilized water (*see* **Note 15**). Incubate the seeds at appropriate temperature and light conditions. Score as alive those seeds that exhibit a protruded radicle that is at least one-third of the size of the seeds.
4. Ensure that the nongerminated seeds were not dormant (*see* **Note 16**).

5. Plot percentages germination against the storage period (days) to create a seed survival curve. This curve exhibits an inverse sigmoidal shape from which a fit can be performed using nonlinear functions (e.g., Hill function) to determine aging rates such as P80 or P50, i.e., the storage time when respectively 80% and 50% of the seed lot remain alive. Alternatively, the survival curve can be modeled using a probit analysis by fitting the viability equation [12] (*see* **Note 17**).

4 Notes

1. The volume of air available above the salt solution should be large enough to accommodate the water vapor that will be released by the seeds, considering that the volume liquid water expands by a factor of 1600 when it turns into vapor.
2. An alternative solution to create an RH at 47% is to dissolve 385 g LiCl to 1 L distilled water [4].
3. If aging is required at 75% RH, a saturated salt solution of NaCl can be used: dissolve the salt in water and keep adding salt until a slurry is formed. Place the slurry in a container and close hermetically. The volume of the slurry should occupy at least 10% of the volume of the container. Let equilibrate for a minimum of 3 days at room temperature.
4. Biological replicates are seed lots harvested from different plants that are grown together. For instance, in Arabidopsis, a total of 6–12 plants of the same genotype can be divided in three biological replicates, with each replicate consisting of a pool of seeds from two to four plants.
5. To obtain a reproducible measure of longevity, seeds should be collected at the point of natural dispersal, when they are considered to be fully mature. Seeds collected too early will not have acquired full longevity (e.g., [13]), and seeds left too long on the mother plant may have reduced viability due to premature aging. Care should be taken that the fruits are allowed to dry out at the end of seed development until the point of seed dispersal. Therefore, air should freely circulate around the plant. For infructescences that show a gradient of development from top to bottom such in Arabidopsis, it is advised to remove the top siliques to increase homogeneity. In addition, all seed batches that will be compared together need to come from the same culture and stored under the same conditions. Broken, shriveled, and empty seeds should be discarded as it will give a false impression of a seed lot exhibiting poor longevity.
6. For each time point during storage, the outcome of the viability assay of one sample will be only an estimate of the viability of

Part III

Biochemical, Molecular Biological and Morphological Analyses

Chapter 7

Analysis of Raffinose Metabolism-Related Sugar Components in Seeds

Tao Li and Tianyong Zhao

Abstract

Raffinose family oligosaccharides (RFOs) are synthesized from sucrose and subsequent addition of galactose moieties which was provided by galactinol. Galactinol is synthesized from UDP-galactose and *myo*-inositol. RFOs accumulate at late stage of seed development and play important roles in seed longevity. RFOs are major components in seeds of many plant species. Here, we document a methodology for extraction and quantitative analysis of raffinose metabolism-related soluble sugars or the derivative alcohols in plant seeds. This protocol, based on high-performance liquid chromatography (HPLC), achieves the efficient separation and accurate quantification of sucrose, *myo*-inositol, galactinol, and raffinose within 25 min of retention time.

Key words Raffinose metabolism, Sugar extraction, High-performance liquid chromatography, Evaporative light-scattering detector, Multicomponent quantitative analysis

1 Introduction

High-performance liquid chromatography (HPLC) is widely used for separation, identification, and quantification of the chemical components in a mixture [1]. Soluble sugars in plant seeds fundamentally act in seed germination, seed dormancy, and seed longevity [2–5]. Raffinose family oligosaccharides (RFOs) accumulate in large amounts during late seed development and control the seed aging tolerance and longevity [5]. HPLC coupled to a variety of detectors, such as refractive index detector (RID) [6], photo-diode array (PAD) [7], and mass spectrum (MS) [8], has been applied for identification and quantitation of plant sugar components. Evaporative light scattering detector (ELSD), as a relative new type of universal detector, is independent of the optical properties and the functional groups of the samples. Any chemical molecule that is less volatile than the mobile phase can be detected by ELSD.

Naoto Kawakami and Kazuhiro Sato (eds.), *Seed Dormancy: Methods and Protocols*, Methods in Molecular Biology, vol. 2830,
https://doi.org/10.1007/978-1-0716-3965-8_7,

This protocol describes a method for detection of the raffinose metabolism components, including sucrose, *myo*-inositol, galactinol, and raffinose in plant seeds. All these sugars and the derivative alcohols can be effectively separated and accurately quantified within a relative shorter time.

2 Materials

All reagents should be at analytical grade, and all the solutions should be prepared using ultrapure water (18 $M\Omega^{-cm}$ at 25 °C). For HPLC, the mobile phase and the standard sugar should be at chromatographically grade.

1. Extraction buffer: 80% ethanol (V/V).
2. Acetonitrile (chromatographically grade).
3. Sucrose, *myo*-inositol, galactinol, and raffinose standards are prepared between 10 and 2000 μg/mL (at least seven different concentration).
4. 0.45 μm membrane filters.
5. Plastic tubes.
6. Waters 1525 pump.
7. Waters 2424 Evaporative Light Scattering Detector.
8. Waters XBridge Amide column.

3 Methods

3.1 Sugar Extraction from Plant Seeds

1. Grind plant seeds (0.5 g dry weight, *see* **Note 1**) into powder in liquid nitrogen using a pestle.
2. Add 3 mL of 80% ethanol (precooling on ice) (*see* **Note 2**) into the powder and homogenate the extraction. Then transfer the homogenates into a clean 15 mL centrifuge tube.
3. Incubate the homogenates in 80 °C water bath for 30 min and mix the extraction upside down for three times during the incubation.
4. Centrifuge the samples at 12,000× *g* for 20 min, and then transfer the supernatants into a new 15 mL centrifuge tube.
5. Extract the remaining sugar in precipitate by another 2 mL of 80% ethanol at 80 °C water bath, centrifuge at 12,000× *g* for 20 min, and combine the supernatant obtained from above extractions (about 7 mL).
6. Remove ethanol by reduced pressure distillation in lyophilizer.

7. Suspend and dissolve the solid in 2 mL distilled water at room temperature for 1 h.
8. Centrifuge the samples at 12,000× *g* (4 °C) for 20 min.
9. Transfer the supernatant into a clean 5 mL centrifuge tube. The samples are stored at −80 °C until HPLC detection (*see* **Note 3**).

3.2 HPLC-ELSD Detection of Sugar Components

1. Install the chromatographic column and guard column following the device connection sequence as shown in Fig. 1 (*see* **Note 4**).
2. De-aerate the mobile phase (H_2O and acetonitrile) by ultrasonic for 30 min. Connect the H_2O and acetonitrile to the two pumps of Waters 1525, respectively.
3. Turn on the power switch of mobile phase pump, ELSD detector, air pump, and column oven in turn.
4. Open the "Breeze 2" software, and connect to the HPLC-ELSD instrument. Set up the parameter of the binary pump to control the flow rate of mobile phase as shown in Table 1. In this method, the flow rate is set to 1 mL min^{-1}.
5. Set up the instrument parameter as following. Column temperature = 30 °C; drift tube temperature = 75 °C; sprayer temperature = 36 °C; sprayer power level = 60%.
6. Open the lamp of ELSD detector, and balance the baseline with a solution containing 65% of acetonitrile and 35% of H_2O for 1 h. After the baseline is stable, perform the sample injection.
7. Take the sugar sample out of −80 °C refrigerator, and melt the sample at room temperature. Filter the samples by 0.45 μm membrane filter (*see* **Note 5**).

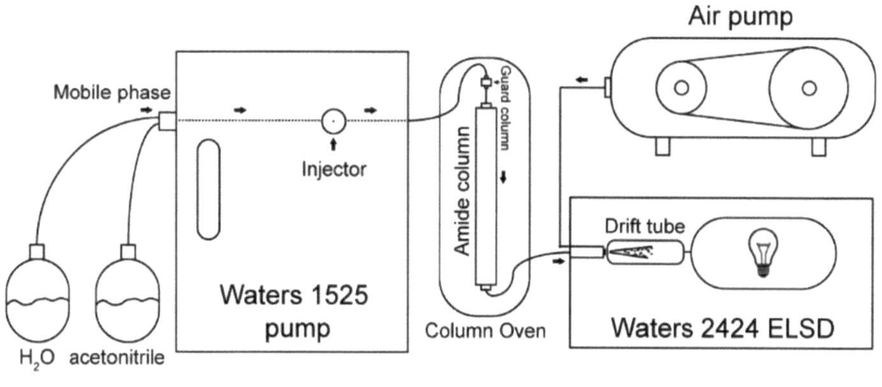

Fig. 1 Connection of column and device in HPLC system. The column should be connected between pumps (Waters 1525) and detector (Waters 2424), and the guard column should be connected to the upstream of the column. The arrows indicated the flow direction of mobile phase or air current

Table 1
Gradient elution procedures

Elution time	Acetonitrile: H_2O ratio
0–1 min	65: 35 change to 70: 30
1–7 min	70: 30 stable
7–8 min	70: 30 change to 63: 37
8–14 min	63: 37 change to 65: 35
14–25 min	65: 35 stable

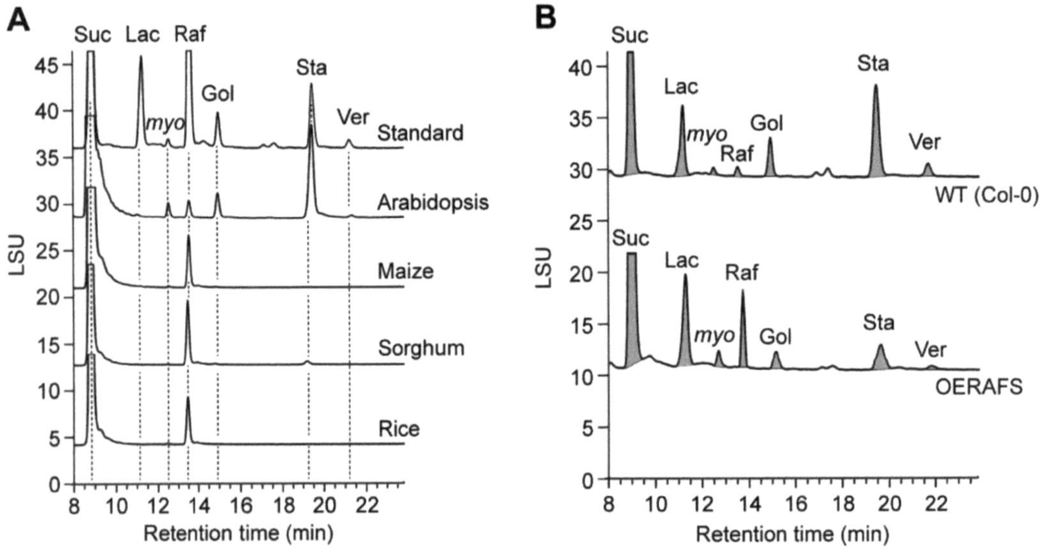

Fig. 2 Examples for qualitative and quantitative analysis of sugars in seed samples. (**a**) Using the chromatographic peak overlap of retention time with sugar standard to identify the seed sugar components of several plant species, the sugar extractions are performed according to the method as described in Subheading 3.1. The overlap of sugar peak is indicated by dotted lines. (**b**) An example to quantify the sugar components in wild-type (WT, Col-0) *Arabidopsis* seeds and *maize raffinose synthase* overexpressing (OERAFS) *Arabidopsis* seeds. After the determination of each sugar components, the area of each peak was calculated by "breeze 2" software. The values of the peak area are used for quantification of the absolute content of the sugar according to Subheading 3.3. The peak area for calculation is indicated by gray color. In this example, lactose was used as an internal reference for correction of the variation created in sugar extraction (see **Note 2**). *Suc* sucrose, *Lac* lactose, *myo myo*-inositol, *Raf* raffinose, *Gol* galactinol, *Sta* stachyose, *Ver* verbascose

8. Click the "injection" button, set up the total retention time at 25 min, and then push the injection port to "loading." Inject the 5–10 μL sample into injection port using micro-syringe, and then push the injection port to "inject" gear (*see* **Note 6**).

9. This gradient elution procedure is able to separate the sugar components related to raffinose metabolism in standard mix and in sugar sample from different plant seeds (Fig. 2a). The different sugar components in seed samples are determined by the overlap with the retention time of standard sugar sample (Fig. 2a). Each sugar peak area of the standard sample and the seed sample should be recorded for the calculation of the sugar content of the seed sample (Fig. 2b). The retention time should be about 9 min for sucrose, 12.5 min for *myo*-inositol, 13.5 min for raffinose, 15 min for galactinol, 19 min for stachyose, and 21 min for verbascose (Fig. 2a).
10. After detection of all samples, close the lamp of ELSD detector, and rinse the column using a methanol gradient elution. Use 70%, 75%, 80%, 85%, 90%, and 95% methanol to wash the column for 20 min each. Then, wash the column with 100% methanol for 30 min and store the column in 100% methanol.
11. Set the drift tube temperature to 30 °C and sprayer to "cooling" mode. After the drift tube temperature drops to below 35 °C, close the instruments.

3.3 Calculations

1. Inject the mixed standard, and read the chromatographic peak area of each sugar components using "breeze 2" software. Take Napierian logarithms of standard concentration and peak area, respectively (*see* **Note 7**). The standard curve can be obtained by fitting the linear equation using the logarithm values. Table 2 and Fig. 1 are examples to show how to obtain the raffinose standard curve.
2. Use the Napierian logarithms of standard concentrations and peak areas to create the standard curve by using EXCEL. You can get an equation of linear regression ($y = ax + b$) (Fig. 3) (*see* **Note 8**). In this equation, "x" indicates Napierian logarithms of standard concentrations (Ln (C)), and "y" indicates Napierian logarithms of peak area of each concentration (Ln (P)).
3. Put the logarithm value of the single sugar peak area into corresponding equation to calculate the logarithm value of the sugar component concentration of the sample. Calculate the seed sugar component content using the formula in Fig. 4.

4 Notes

1. For different plant seeds, the sample weight for sugar extraction should be optimized. In our lab, we used 0.5 g for *Arabidopsis* seeds, 0.5 g for maize embryos, and 2 g for maize endosperms. The plant seeds should be dried by using high-temperature drying oven (180 °C for 5 min, following by 80 °C for drying to constant weight).

Table 2
Raffinose standard curve value calculation

Concentration (μg/mL)[a]	10	20	40	60	80	100	200	400	600	800	1000	2000
Peak area[b]	3524	10,176	30,232	54,325	82,091	118,358	306,881	869,829	1,948,872	2,979,202	4,268,478	13,656,529
Ln (C)[c]	2.30	3.00	3.69	4.09	4.38	4.61	5.30	5.99	6.40	6.68	6.91	7.60
Ln (P)[d]	8.17	9.23	10.32	10.90	11.32	11.68	12.63	13.68	14.48	14.91	15.27	16.43

[a]Raffinose standard concentrations (10–2000 μg/mL)
[b]Raffinose peak area calculated by "breeze 2" software
[c]Napierian logarithms of each Raffinose standard concentration are used to construct standard curve
[d]Napierian logarithms of peak area of each standard concentrations are used to construct standard curve

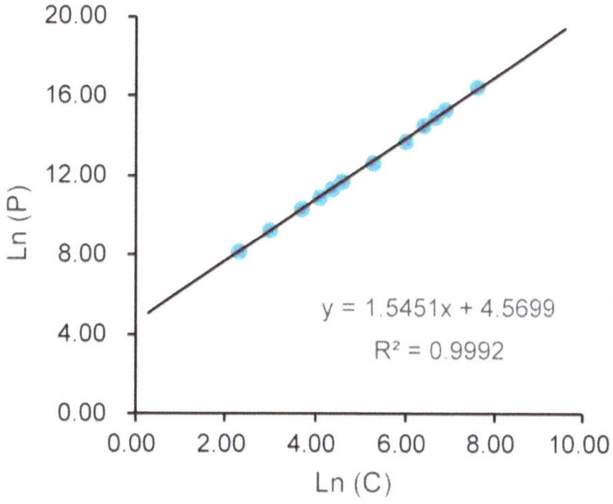

Fig. 3 Establishment of standard curve and the corresponding regression equation based on the logarithms values from Table 2. The values in the *X* axis indicate the Ln (C) of Table 2, and the values in the *Y* axis indicate the Ln (P) of Table 2. The line in the figure indicates the fitting linear curve. The corresponding equation is shown as $y = 1.5451x + 4.5699$. R^2 is the correlation coefficient

$$\text{Sugar content} = \frac{e^{\text{Ln (C)}} \times V_{\text{Extracion}}}{\text{MW} \times W_{\text{sample}}}$$

Fig. 4 Calculation formula of sugar component content in seed sample. "e" indicates natural exponential, "$V_{\text{Extraction}}$" indicates the volume of total extraction (in this protocol, it is 7 mL), "*MW*" indicates relative molecular mass of each sugar component, and "W_{Sample}" indicates dry weight of the seed sample. The sugar content unit is μmol mg^{-1} dry weight

2. Since lactose does not exist in plant seed, a certain amount of lactose (200 μg/mL is recommended) can be added to the 80% ethanol as the external standard. The variation caused by the extraction process can be eliminated by referring to the loss of lactose in the final calculation.

3. Before HPLC testing, the sample must be filtered through a 0.45 μm filter to prevent clogging of the column. Furthermore, one can choose to remove the salt in samples by using Sephadex G-25 column.

4. Please install the chromatographic column following the instruction, and special attention should be paid to the direction of the column installation.

5. It is more convenient to use an automatic injector. However, if one chooses a manual injector, please clean the injection hole using methanol before next injection to avoid contamination from previous sample.

6. Please dilute your sample if necessary to ensure the signal abundance is within the standard curve.
7. Unlike other detectors, signal captured by ELSD detector is exponential in relation to the sample concentration. Thus, one needs to use the logarithm of signal values and standard sample concentrations to obtain the linear standard curve.
8. Individual standard curves are required for each sugar component. Mixed standard sample containing all sugar components can be used to obtain different sugar curves. In addition, the sugar composition in the sample is characterized according to the overlap of retention time with standard.

References

1. Blum F (2014) High performance liquid chromatography. Br J Hosp Med (Lond, Engl 2005) 75(2):C18–C21. https://doi.org/10.12968/hmed.2014.75.Sup2.C18
2. Wang Y, Cui Y, Hu G, Wang X, Chen H, Shi Q, Xiang J, Zhang Y, Zhu D, Zhang Y (2018) Reduced bioactive gibberellin content in rice seeds under low temperature leads to decreased sugar consumption and low seed germination rates. Plant Physiol Biochem 133:1–10. https://doi.org/10.1016/j.plaphy.2018.10.020
3. Leprince O, Pellizzaro A, Berriri S, Buitink J (2017) Late seed maturation: drying without dying. J Exp Bot 68(4):827–841. https://doi.org/10.1093/jxb/erw363
4. de Souza Vidigal D, Willems L, van Arkel J, Dekkers BJW, Hilhorst HWM, Bentsink L (2016) Galactinol as marker for seed longevity. Plant Sci 246:112–118. https://doi.org/10.1016/j.plantsci.2016.02.015
5. Li T, Zhang Y, Wang D, Liu Y, Dirk LMA, Goodman J, Downie AB, Wang J, Wang G, Zhao T (2017) Regulation of seed vigor by manipulation of raffinose family oligosaccharides in maize and *Arabidopsis thaliana*. Mol Plant 10(12):1540–1555. https://doi.org/10.1016/j.molp.2017.10.014
6. Tsung Min Kuo JFV, Wolf WJ (1988) Content of raffinose oligosaccharides and sucrose in various plant seeds. J Agr Food Chem 36:32–36
7. Schneider T, Keller F (2009) Raffinose in chloroplasts is synthesized in the cytosol and transported across the chloroplast envelope. Plant Cell Physiol 50(12):2174–2182. https://doi.org/10.1093/pcp/pcp151
8. Georgelis N, Fencil K, Richael CM (2018) Validation of a rapid and sensitive HPLC/MS method for measuring sucrose, fructose and glucose in plant tissues. Food Chem 262:191–198. https://doi.org/10.1016/j.foodchem.2018.04.051

Chapter 8

Chromatin Immunoprecipitation Using Imbibed Seeds of *Arabidopsis thaliana*

Yongmin Kwon and Giltsu Choi

Abstract

Chromatin immunoprecipitation (ChIP) is used to analyze the targeting of a protein to a specific region of chromatin in vivo. Here, we present an instructive ChIP protocol for Arabidopsis imbibed seeds. The protocol covers all steps, from the sampling of imbibed seeds to the reverse crosslinking of immunoprecipitated protein–DNA complexes, and includes experimental tips and notes. The targeting of the protein to DNA is determined by quantitative PCR (qPCR) using reverse crosslinked DNA. The protocol can be further scaled up for ChIP-sequencing (ChIP-seq) analysis. As an example of the protocol, we include a ChIP-quantitative PCR (ChIP-qPCR) analysis demonstrating the targeting of PIF1 to the *ABI5* promoter.

Key words Chromatin immunoprecipitation, ChIP, ChIP-qPCR, ChIP-seq, *Arabidopsis thaliana*, Seed, DNA binding, Promoter targeting

1 Introduction

Chromatin immunoprecipitation (ChIP) is a procedure used to examine the targeting of a protein to specific regions of chromatin in vivo [1]. The procedure comprises a series of steps, including the crosslinking of the protein to chromatin in vivo, the fragmentation of chromosomal DNA to small sizes, and the immunoprecipitation of protein–chromosomal DNA complexes with an antibody against the protein. The coimmunoprecipitated DNA fragments can be subject to quantitative PCR amplification for a specific DNA sequence (ChIP-qPCR) or direct sequencing of all precipitated DNA fragments (ChIP-seq).

The ChIP methodology was pioneered by David Gilmour and John T. Lis in a study that used UV irradiation-based crosslinking to examine the association of RNA polymerase with DNA [2]. Subsequent studies used formaldehyde as the crosslinker to increase the crosslinking-arm distance between proteins and DNA [3], or applied double crosslinking with disuccinimidyl glutarate

Naoto Kawakami and Kazuhiro Sato (eds.), *Seed Dormancy: Methods and Protocols*, Methods in Molecular Biology, vol. 2830, https://doi.org/10.1007/978-1-0716-3965-8_8,

(DSG) and formaldehyde to detect indirect interactions between proteins and chromatin [4]. Quantitative PCR (qPCR) was adopted to quantify specific immunoprecipitated DNA fragments (ChIP-qPCR) [5], and next-generation sequencing (NGS) was applied to determine the genome-wide targeting sites of a protein (ChIP-seq) [6]. To date, many variations of the ChIP assay have been developed [7–9].

ChIP assays have been successfully used to study protein–DNA interactions in plants and various plant parts, including whole seedlings [10], mesophyll protoplasts [11], inflorescence stem [12], and seeds [13]. Here, we describe a detailed ChIP method for Arabidopsis imbibed seed samples, with experimental tips provided for each step. The protocol also includes notes on scaling up the strategy for ChIP-seq.

2 Materials

Prepare the following solutions using Milli-Q quality ultrapure water (18 MΩ/cm at 25 °C), except for the MS agar medium, which can be made with distilled water. Use analytic grade reagents.

2.1 Plant Materials

1. Arabidopsis seeds (*see* **Note 1**).

2.2 Reagents

1. Seed sterilization solution: 0.015% (v/v) Triton-X 100, 10% (v/v) bleach in double-distilled water (DDW).
2. 10× PBS (crosslinking solution): 1.37 M NaCl, 27 mM KCl, 100 mM disodium monohydrogen phosphate, 18 mM mono-potassium dihydrogen phosphate (*see* **Note 2**).
3. 2 M glycine solution.
4. ChIP buffer: 50 mM HEPES, 150 mM NaCl, 1 mM EDTA, 0.1% (w/v) sodium deoxycholate, 1% (v/v) Triton X-100, 100 μM MG132, 1 mM PMSF, protease inhibitor cocktail (*see* **Note 3**).
5. Low-salt wash buffer: 20 mM Tris, 2 mM EDTA, 0.5% (v/v) Triton X-100, 0.2% (w/v) SDS, 150 mM NaCl, HCl-titrated pH 8.00.
6. High-salt wash buffer: 20 mM Tris, 2 mM EDTA, 0.5% (v/v) Triton X-100, 0.2% (w/v) SDS, 500 mM NaCl, HCl-titrated pH 8.00.
7. LiCl wash buffer: 10 mM Tris, 1 mM EDTA, 0.5% (w/v) sodium deoxycholate, 0.5% (v/v) NP-40, 250 mM LiCl, HCl-titrated pH 8.00.
8. TE buffer: 10 mM Tris, 1 mM EDTA, HCl-titrated pH 8.00.

9. Reverse crosslinking buffer: 10 mM Tris, 1 mM EDTA, 1% (w/v) SDS, 300 mM NaCl, HCl-titrated pH 8.00.
10. RNase A.
11. Proteinase K.
12. Silica type column-based DNA purification kit.
13. Antibody solution.

3 Methods

3.1 Preparation of Imbibed Seed Samples

1. Sterilize the dry seeds with 10 mL of the seed sterilization solution in a 15-mL Falcon tube by vigorous shaking for 7 min (60 mg per biological replicate for ChIP-qPCR analysis; 400 mg per biological replicate for ChIP-seq) (*see* **Note 4**).
2. Immediately wash seeds with 10 mL of DDW with a few vigorous shakes. When seeds are sedimented to the bottom of the tube, remove the water using vacuum suction. Repeat the washing procedure five more times (*see* **Note 5**).
3. Spread seeds evenly on two layers of 3 M filter paper laid on half-strength MS-agar plates. For chemical treatment, use half-strength MS-agar plates containing the chemical.
4. Incubate seeds at 22 °C in a growth chamber for the desired time with or without pretreatment (*see* **Note 6**).

3.2 Fixation of Chromatin Complexes with Formaldehyde

1. After the incubation step, collect all seeds on the filter paper and put the seeds in 10 mL of 1× PBS (see the Materials section) in 50 mL Falcon tube for each biosample. Make sure that all seeds are immersed in the solution.
2. Add 277.8 μL of 37% formaldehyde solution (final concentration, 1%) to each tube (*see* **Note 7**).
3. Place each tube in a vacuum jar and apply vacuum at 720 mmHg for 1 h. While vacuuming, prepare the 2 M glycine solution.
4. Slowly reduce the pressure to 0 mmHg at a rate of −6 mmHg/s.
5. After the vacuum treatment, quench free formaldehydes by adding 500 μL of 2 M glycine directly into the tube. Mix several times by inverting and let the tube sit for 5 min (*see* **Note 8**).
6. Wash the seeds with Milli-Q quality water using a Büchner funnel under a mild vacuum treatment (300 mmHg). Put one layer of nylon mesh on the top of the 3 M filter paper on the Büchner funnel. To wash, pour seeds onto the nylon mesh layer and flow 50 mL of water directly over seeds. Then, apply

the vacuum for 30 more seconds to remove excess water (*see* **Note 9**).

7. Using a sterile stainless steel spatula, collect all seeds on the nylon mesh into a 2-mL safe-lock tube with a 6-mm titanium bead and flash-freeze in liquid nitrogen (*see* **Note 10**). The frozen samples can be stored in a deep-fridge (−80 °C) until the grinding step.

3.3 Grinding of Seed Samples with the TissueLyser

1. Set the shaking frequency of the TissueLyser to 23 Hz and the duration to 1 min and 30 s (*see* **Note 11**).
2. Chill a TissueLyser adapter set 2 × 24 by dipping the set in liquid nitrogen.
3. Take out the adapter set and place the frozen seed sample tubes in the set.
4. Install the adapter set in the TissueLyser and grind.
5. Uninstall the adapter set and chill the set with the sample tubes by dipping them in liquid nitrogen.
6. Repeat **steps 4** and **5** twice more.
7. The seeds should have become a fine homogenous powder (Fig. 1). If not, perform additional rounds of grinding.

3.4 Isolation of Nuclei

1. Resuspend the sample powder in 2 mL of ChIP buffer by pipetting.
2. Filter through two layers of Miracloth under 3500 *g* at 4 °C for 20 min in a 15-mL Falcon tube (*see* **Note 12**).
3. Discard the supernatant and resuspend the nuclear pellet fraction in 1.7 mL of ChIP buffer with 0.1% (w/v) SDS, which facilitates the lysis of nuclei.

3.5 Shearing of Chromatin

1. Set the Bioruptor at the high-power output for a 30 s on|30 s off cycle. Fill the chamber with ice-cold water.
2. Apply five cycles of sonic pulses to the resuspended nuclei using the Bioruptor.
3. Replace the water with fresh ice-cold water.
4. Repeat **steps 2** and **3** four more times (total, 25 cycles) to shear the DNA to fragments of 200–500 bps for ChIP-qPCR or six more times (total, 35 cycles) to shear the DNA to fragments of 100–300 bps for ChIP-seq analysis (Fig. 2) (*see* **Note 13**).
5. Transfer the sonicated DNA to a 1.7-mL microfuge tube. Remove debris by centrifugation (15,000 *g*) at 4 °C for 5 min using a temperature-controlled high-speed tabletop centrifuge, and then transfer all of the supernatant to a new 1.7-mL microfuge tube. Centrifuge the new tube as described above,

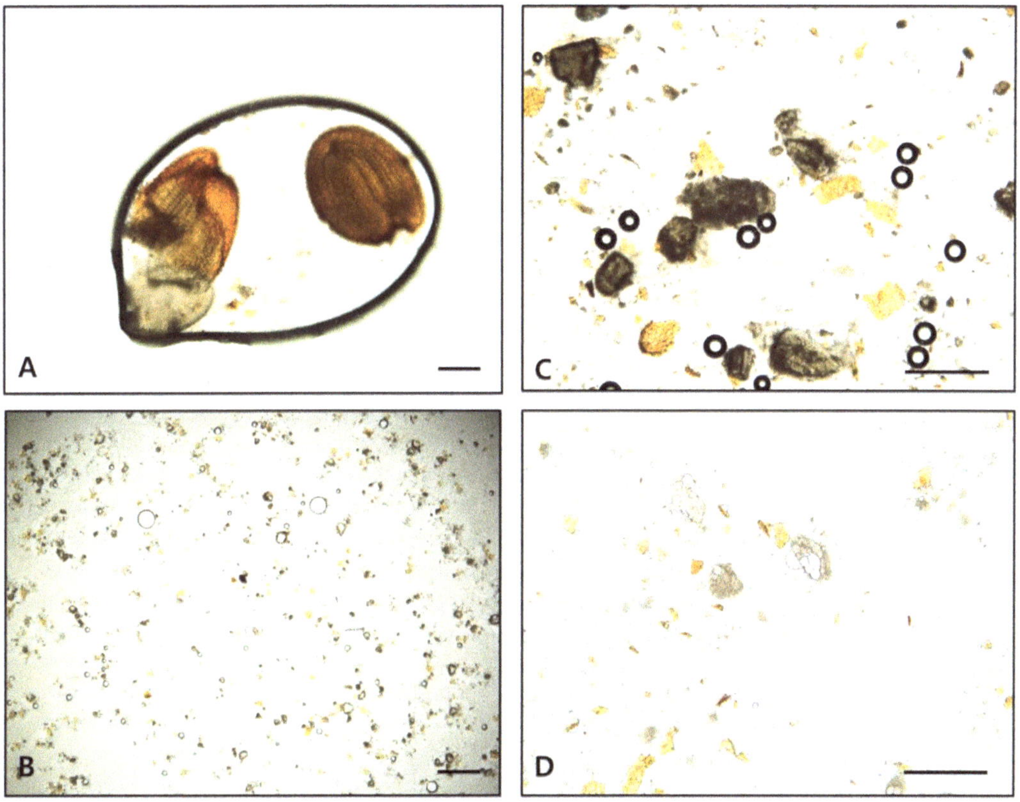

Fig. 1 The grinding of crosslinked frozen seeds with a TissueLyser. Crosslinked seeds are collected in a 2-mL safe-lock tube containing a titanium bead of different diameters, frozen by dipping the tube in liquid nitrogen, and ground using a TissueLyser for three rounds. The shaking frequency of TissueLyser is set to 23 Hz and the duration to 1 min and 30 s. The ground powder is observed under a light microscope. Seeds are ground with a 3-mm titanium bead (**a**, magnification ×100; bar = 100 μm, **c**, ×200; bar = 100 μm) or a 6-mm titanium bead (**b**, magnification ×100; bar = 100 μm, **d**, ×200; bar = 100 μm). The 6-mm titanium bead yields a more homogenous powder

and transfer 1.3 mL of the supernatant to a new 1.7-mL microfuge tube (*see* **Note 14**).

6. Set aside 60 μL of the sample (5% fraction of the immunoprecipitation sample) as the input control; flash-freeze in liquid nitrogen and store in a deep-fridge until the reverse crosslinking step.

3.6 Immunoprecipitation of the Protein–Chromatin DNA Complexes

1. Transfer 1.2 mL of sample to a new tube for immunoprecipitation. Add 100 ng of monoclonal antibody to the 1.2 mL of sample and incubate on a rotary shaker (12 rpm) at 4 °C for 12 to 17 h (*see* **Note 15**).
2. Add 30 μL of ChIP buffer-equilibrated protein A agarose or protein G agarose resin to the sample and incubate on a rotary shaker (12 rpm) at 4 °C for 2 h (*see* **Note 16**).

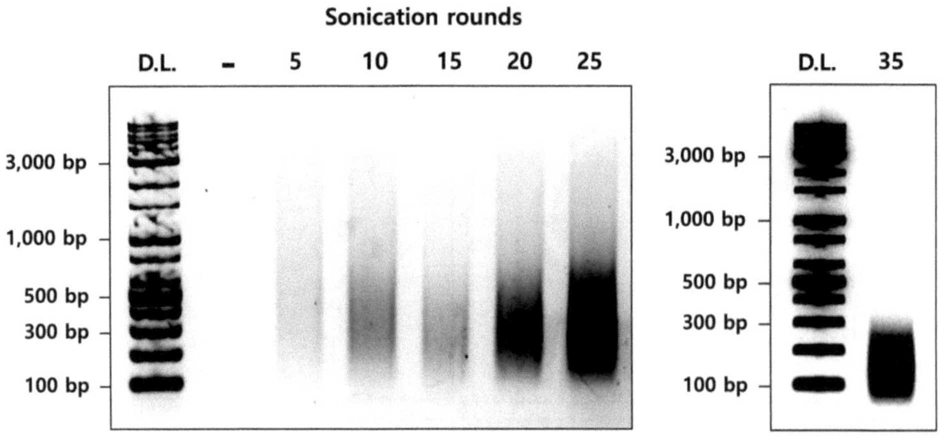

Fig. 2 The shearing of chromatin DNA with successive sonication cycles. The two stained gels show the degree of chromatin DNA fragmentation according to the rounds of sonication. The DNA samples are purified with a PCR purification kit after reverse-crosslinking (refer to the reverse-crosslinking steps.). The unsheared chromosomal DNA is not purified efficiently due to its extreme length

3. Precipitate the resin by centrifugation (500 *g*) at 4 °C for 2 min and remove the supernatant carefully to minimize the loss of resin.

3.7 Washing and Elution of Immunoprecipitated Protein–Chromatin DNA Complexes

1. Add 1000 μL of low-salt wash buffer to the resin and wash on a rotary shaker (12 rpm) at 4 °C for 5 min. Precipitate the resin by centrifugation (500 *g*) at 4 °C for 2 min and remove the supernatant carefully to minimize the loss of resin. Repeat the washing step a second time. The resin sedimentation should be handled carefully with minimal loss. Follow the wash steps below (*see* **Note 17**).
2. Wash the beads two times with the high-salt wash buffer.
3. Wash the beads two times with LiCl wash buffer.
4. Wash the beads two times with TE wash buffer.
5. Resuspend the resin in 200 μL of reverse crosslinking buffer with gentle tapping. Also add 140 μL of reverse crosslinking buffer to the deep-fridge-kept 5% input sample. The following steps apply to both samples. Tightly seal the tubes with parafilm and incubate them at 65 °C for 12 h (*see* **Note 18**).
6. Treat the reverse crosslinked samples with a final concentration of 100 ng/mL RNase A (commercial DNase-free RNase A or heat-treated lab-made RNase A). Incubate the samples at 37 °C for 30 min (*see* **Note 19**).
7. Add proteinase K at a final concentration of 50 μg/mL directly to the RNA-digested samples. Incubate the samples at 45 °C for 2 h.

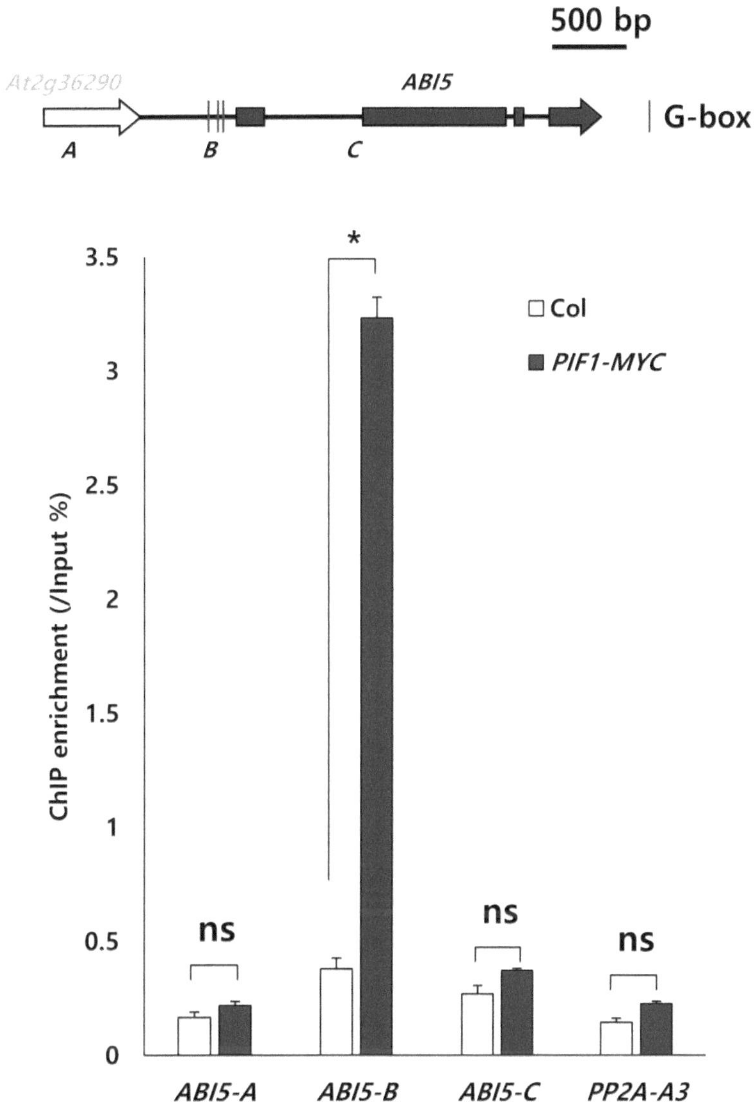

Fig. 3 The targeting of PIF1 to the *ABI5* promoter, as determined by ChIP-qPCR. ChIP assay was with wild-type (Col) and transgenic seeds expressing MYC-tagged PIF1 (*PIF1-MYC*) and MYC antibody. The upper diagram indicates an *ABI5* genomic region with G-boxes (vertical bars) and ChIP amplicons (*A*, *B*, *C*). *PP2A-A3* is a coding region of *PP2A-A3* gene, which was used as a nonbinding control region (*see* **Note 21**). Star marks indicate significant enrichment ($n = 3$ biological replicates; *t*-test, * $p < 0.05$, *ns* not significant)

8. Precipitate the resin by centrifugation (500 *g*) at room temperature for 2 min, and transfer the supernatant to a new microfuge tube. Purify the DNA using a PCR purification kit (e.g., Qiagen PCR purification kit). Elute twice with 100 μL of DDW.

9. The eluted DNAs are now ready for the quantitative PCR analysis. Using this protocol, the representative data of the transcription factor PIF1, targeting onto the *ABI5* promoter [14], containing G-boxes, are shown in Fig. 3. The input sample quantitation will correspond to 5% of the total fraction (*see* **Note 20**).

4 Notes

1. Before performing the ChIP assay, we recommend that a germination assay be performed to ensure the seed quality and that the germination frequencies are reproducible under the adopted experimental conditions.
2. We do not recommend autoclave sterilization of 10× stock solution, as this causes precipitation of phosphate. The 10× PBS stock solution is relatively safe from contamination because of its high salt concentration. We recommend filter sterilization using a 0.2-μm syringe filter when preparing the 1× PBS. The pH of the 1× PBS should be between pH 7.30 and pH 7.50 (usually pH 7.42). Note that the pH of the 10× stock solution will be slightly acidic, since the phosphate buffer is dilution-sensitive.
3. To sterilize the ChIP buffer and the wash buffers, use filter sterilization (0.2 μm).
4. We recommend that three biological replicates of the two seed samples being compared should be processed simultaneously. Do not shake more than 10 min, as this can damage the seeds.
5. Any loss of seeds during the removal of water will cause sample-to-sample variation; thus, pay attention not to losing any seed.
6. The seed preparation steps may be adjusted to the experimental purpose. For any condition or treatment, however, we recommend that the seeds not be incubated in a growth chamber for more than 12 h unless there is a specific reason to do so. Under phyB-dependent germination conditions, germination becomes irreversible at 12–15 h after the red-light pulse, and the testa begins to rupture at 24 h after the red-light pulse. If seeds are cold-imbibed, germination becomes irreversible at 9–12 h. Thus, the DNA-binding patterns and gene expression patterns observed at times falling after the induction of germination are likely to reflect different developmental stages. To compare the targeting of a protein under a phyB-dependent seed germination condition, irradiate seeds with a far-red light pulse (2.59 μmol m^{-2} s^{-1}, phyBoff condition) or a far-red light pulse followed by a red pulse (11.5 μmol m^{-2} s^{-1}, phyBon condition). Then wrap the plates with aluminum foil in the

dark (i.e., under a green safety light) and incubate them at 22 °C in a growth chamber for the desired time. For a phyA-dependent seed germination condition, irradiate seeds with a far-red light pulse (2.59 μmol m^{-2} s^{-1}), wrap plates with aluminum foil in the dark (green safety light), and incubate at 4 °C for 48 h. For the phyAon condition, remove the aluminum foil, irradiate the seeds with far-red light, rewrap the plates with aluminum foil in the dark (green safety light), and incubate them at 22 °C in a growth chamber for the desired time. For the phyAoff condition, transfer directly to the growth chamber after the 48-h cold-imbibition step. We typically use three sheets of aluminum foil to block light from penetrating via a crack or accidental tear.

7. Do not use any reactive-amine-based buffer (e.g., Tris, glycine, histidine, HEPES, MOPS, or MES), as these buffering agents react with formaldehyde, and reduce the crosslinking efficiency. We recommend using a fresh formaldehyde solution (37%).
8. If a specific light condition is required, the collection and crosslinking steps should be done under the specific light condition. For the dark condition, use a dim green-light condition.
9. The nylon mesh helps collect the seeds after washing, while the filter paper absorbs the excess water.
10. Spread the seeds on the tube wall before freezing, as large seed clumps are difficult to grind. If more than 60 mg of dry seeds are needed for a sample, we recommend they be collected in two or more tubes. We have examined two sizes of the titanium beads (the diameters with 3 mm or 6 mm), and the optimal size of the bead was 6 mm for obtaining more homogenous powder from the seed sample (Fig. 1).
11. Establish the setting and cycle number for your own equipment. For a ChIP-seq assay that uses more than 400 mg of dry seeds per sample, we recommend that a liquid nitrogen-chilled mortar and pestle be used for grinding. Do not allow the samples to melt during grinding. Wear safety gloves.
12. Put two layers of Miracloth in a 15-mL Falcon tube, fitted into a concave shape that will suffice to hold the resuspended sample. After pouring the resuspended sample, close the cap tightly before the centrifugation.
13. Establish the setting and cycle number for your own equipment that will shear the DNA to 200–500 bps fragments for ChIP-qPCR or 100–300 bps fragments for ChIP-seq.
14. We found that removing debris twice helps reduce the noise signals in ChIP-qPCR and ChIP-seq.
15. The antibody quality is critical for a successful ChIP assay; thus, a ChIP-grade antibody corresponding to an epitope-tag or a

specific target protein should be used. Laboratory-made antibodies that can successfully detect denatured proteins should be tested for their suitability for a ChIP assay. The necessary amount of antibody should be titrated empirically for each antibody. Generally, 100 ng of the monoclonal IgG is sufficient for ChIP-qPCR.

16. We recommend using salmon sperm DNA-coated protein A agarose resin, to prevent nonspecific binding.
17. For 30 μL of resin, 1000 μL of wash buffer is sufficient. For more resin (e.g., in a ChIP-seq experiment), scale up the wash buffer volume according the resin volume.
18. The use of a high-salt buffer condition (e.g., 300 mM NaCl) and strong surfactants (e.g., SDS) will help solubilize the precipitated immune complexes. We recommend using at least 300 mM of salt and 1% (w/v) of SDS. The percentage of the free DNA and protein complex per time (reverse crosslinking rate) depends on temperature in an exponential manner. For example, 13.75% of the reversed complex can be obtained in 1 h at 65 °C, but only 3.89% obtained in 1 h at 45 °C [15]. The reverse crosslinking buffer should be made using a sufficient amount of an amine-based buffering agent (e.g., 10 mM of Tris). The reversed formaldehyde is easily quenched by Tris.
19. It is critical that the abundant RNA be removed from the lysate. Since the molecular nature of DNA and RNA is similar, RNA interferes with many purification methods, including column-type purification and alcohol precipitation.
20. We recommend that a novice practice at least twice with a set of wild-type and transgenic seeds that express a high level of MYC- or other tagged DNA-binding transcription factor has been previously used for ChIP-qPCR analysis. If data are obtained reliably with the practice set, proceed with experimental sets.
21. The primers used in Fig. 3 are following. The qPCR primers for the amplicon A in the *ABI5* genomic region are 5′-TTCC CGTTTGCTCCAGGTATG-3′ (forward primer) and 5′-TCT CATGAATCTCTCTGATCCCGG-3′ (reverse primer), for the amplicon B in the *ABI5* genomic region are 5′-CATTTGTG TAGCCGAAGTCAC-3′ (forward primer) and 5′-GGAATGC TAACTTAGACAGGTCC-3′ (reverse primer), and primers for the amplicon C in the *ABI5* genomic region are 5′-TTGTTA TTGTCGTTTAGGTGGTATCC-3′ (forward primer) and 5′-GCCATGGACGACTCTACTTCTC-3′ (reverse primer). For the amplicon of the nontarget control region (*PP2A-A3*), the forward primer is 5′-CATTGTAGAACTTGCTGAAGACAGG -3′ and the reverse primer is 5′- CAAGACGCTTCA GATTGTTTGCAGC-3′.

References

1. Orlando V (2000) Mapping chromosomal proteins in vivo by formaldehyde-crosslinked-chromatin immunoprecipitation. Trends Biochem Sci 25(3):99–104
2. Gilmour DS, Lis JT (1984) Detecting protein-DNA interactions *in vivo*: distribution of RNA polymerase on specific bacterial genes. Proc Natl Acad Sci USA 81(14):4275–4279
3. Solomon MJ, Larsen PL, Varshavsky A (1988) Mapping protein DNA interactions in vivo with formaldehyde: evidence that histone H4 is retained on a highly transcribed gene. Cell 53(6):937–947
4. Wang Z, Zang C, Cui K, Barski A, Peng W, Zhao K (2009) Genome-wide mapping of HATs and HDACs reveals distinct functions in active and inactive genes. Cell 138(5): 1019–1031
5. Mukhopadhyay A, Deplancke B, Walhout AJ, Tissenbaum HA (2008) Chromatin immunoprecipitation (ChIP) coupled to detection by quantitative real-time PCR to study transcription factor binding to DNA in Caenorhabditis elegans. Nat Protoc 3(4):698–709
6. Johnson DS, Mortazavi A, Myers RM, Wold B (2007) Genome-wide mapping of in vivo protein-DNA interactions. Science, pp 1497–1502
7. O'Neill L, VerMilyea M, Turner B (2006) Epigenetic characterization of the early embryo with a chromatin immunoprecipitation protocol applicable to small cell populations. Nat Genet 38:835–841
8. Rhee HS, Pugh BF (2011) Comprehensive genome-wide protein-DNA interactions detected at single-nucleotide resolution. Cell 147(6):1408–1419
9. Song L, Crawford GE (2010) DNase-seq: a high-resolution technique for mapping active gene regulatory elements across the genome from mammalian cells. Cold Spring Harb Protoc, p prot5384. https://doi.org/10.1101/pdb.prot5384
10. Sakuraba Y, Jeong J, Kang M-Y, Kim J, Paek N-C, Choi G (2014) Phytochrome-interacting transcription factors PIF4 and PIF5 induce leaf senescence in Arabidopsis. Nat Commun 5: 4636
11. Lee JH, Jin S, Kim SY, Kim W, Ahn JH (2017) A fast, efficient chromatin immunoprecipitation method for studying protein-DNA binding in Arabidopsis mesophyll protoplasts. Plant Methods 13:42
12. Yamaguchi N, Winter CM, Wu MF, Kwon CS, William DA, Wagner D (2014) PROTOCOLS: chromatin immunoprecipitation from Arabidopsis tissues. Arabidopsis Book 12: e0170
13. Oh E, Yamaguchi S, Hu J, Jikumaru Y, Jung B, Paik I, Lee H-S, Sun T-p, Kamiya Y, Choi G (2007) PIL5, a phytochrome-interacting bHLH protein, regulates gibberellin responsiveness by binding directly to the GAI and RGA promoters in Arabidopsis seeds. Plant Cell 19(4):1192–1208
14. Oh E, Kang H, Yamaguchi S, Park J, Lee D, Kamiya Y, Choi G (2009) Genome-wide analysis of genes targeted by PHYTOCHROME INTERACTING FACTOR 3-LIKE5 during seed germination in Arabidopsis. Plant Cell 21(2):403–419
15. Kennedy-Darling J, Smith LM (2014) Measuring the formaldehyde protein-DNA cross-link reversal rate. Anal Chem 86(12):5678–5681

Chapter 9

Confocal Imaging of Seeds

Dennys-Marcela Gómez-Páez and Enrico Magnani

Abstract

In flowering plants, proper seed development is achieved through the constant interplay of fertilization products, embryo and endosperm, and maternal tissues. Understanding such a complex biological process requires microscopy techniques able to unveil the seed internal morphological structure. Seed thickness and relatively low permeability make conventional tissue staining techniques impractical unless combined with time-consuming dissecting methods. Here, we describe two techniques to imaging the three-dimensional structure of *Arabidopsis* seeds by confocal laser scanning microscopy. Both procedures, while differing in their time of execution and resolution, are based on cell wall staining of seed tissues with fluorescent dyes.

Key words *Arabidopsis*, Seed development, Seed architecture, Cell wall, Starch, Staining, Propidium iodide, SR2200, Calcofluor, Confocal laser scanning microscopy

1 Introduction

Confocal laser scanning microscopy (CLSM) has been widely used to obtain three-dimensional (3D) images of fluorescent cellular groups without requiring tissue sectioning. In the absence of auto-fluorescence, tissues can be stained with fluorescent dyes for their visualization. Propidium iodide (PI), a fluorescent dye able to bind DNA, cell walls, and starch, has been used in multiple applications such as measuring apoptosis [1–3], sorting cells based on DNA content [4, 5], and nuclei staining of dead cells [6]. In *Arabidopsis*, CLSM imaging of tissues stained with PI has been successfully employed for cell wall analyses [7, 8], starch detection [9, 10], and morphological studies of different tissues [7, 11–15].

Based on the conventional pseudo-Schiff propidium iodide (PS-PI) staining technique, used to observe by CLSM the cellular architecture of several *Arabidopsis* tissues [13–15], Truernit et al. [7] developed a modified pseudo-Schiff propidium iodide (mPS-PI) staining method to obtain higher-resolution images, perform 3D reconstructions of plant tissues, and study their cellular organization. This latter technique has been applied to the analysis

Naoto Kawakami and Kazuhiro Sato (eds.), *Seed Dormancy: Methods and Protocols*, Methods in Molecular Biology, vol. 2830, https://doi.org/10.1007/978-1-0716-3965-8_9,

of different plant organs to characterize their biological functions and monitor their development [7, 11, 12].

As an alternative to PI staining, which is not specific for the cell wall, Renaissance SR2200 and calcofluor White M2R staining have been implemented [16, 17]. These dyes, which specifically stain beta-1,3- and beta-1,4-glucans in plant cell walls, thus allowing to highlight plant cell outlines, have been used to morphologically characterize reproductive organs [16] and monitor ovule development in *Arabidopsis* [18].

In this chapter, we provide two step-by-step methods used to investigate the cellular structure of *Arabidopsis* seeds based on PI and Renaissance SR2200 reagent staining, followed by CLSM imaging.

2 Materials

2.1 Harvesting of Siliques

1. Three to four-week-old *Arabidopsis* plants.
2. Soft thread or micropore tape.
3. Magnifying glass.
4. Laboratory dissecting scissors.
5. Two high-precision micro tweezers.
6. Thin syringe needle (Preferably 26G × 0.5″; 0.45 × 12 mm).
7. Microscope slides.
8. Double-sided tape.
9. Stereo microscope.
10. Tissue culture well plates with flat bottom.
11. Medium incubation baskets.
12. Milli-Q or nonpyrogenic sterile water.
13. 1000 μL adjustable-volume micropipette.
14. Disposable pipette tips (1000 μL volume).
15. Parafilm M.

2.2 Modified Pseudo-Schiff Propidium Iodide (mPS-PI) Staining Method Adapted to the Analysis of Arabidopsis *Seed Cellular Structures*

1. Fixative solution: GreenFix; a commercial ethanedial and ethyl alcohol aqueous solution Glyoxal-based formalin fixative substitute) [19] (*see* **Note 1**).
2. NaOH/SDS buffer: 0.2 N NaOH and 1% (w/v) SDS (*see* **Note 2**).
3. Starch digestion buffer: 0.1 M $NaPO_4$, pH 6.9.
4. Starch digestion solution: 0.4 mg/mL α-amylase from Bacillus sp. in 20 mM $NaPO_4$ and 6.7 mM NaCl solution, pH 6.9 (*see* **Note 3**).

5. Bleaching solution: 1.25% (v/v) sodium hypochlorite (NaClO) in water.
6. Aldehyde oxidizing solution: 1% (v/v) periodic acid (H_5IO_6) in water from a 50% stock solution (stored at 4 °C).
7. Pseudo-Schiff (PS) solution (80%): 0.019 g/mL sodium bisulfite ($NaHSO_3$) in 0.15 M HCl (work under a chemical fume hood).
8. Propidium iodide (PI): 2 mg/mL PI in water (work under a chemical fume hood). Store at 4 °C protected from light.
9. Chloral hydrate mounting buffer: 4 g/mL chloral hydrate in a 33% (v/v) glycerol solution (work under a chemical fume hood).
10. Chemical fume hood.
11. Stereo microscope.
12. Milli-Q or nonpyrogenic sterile water.
13. Microscope slides.
14. Rectangular glass coverslips, 22 × 50 mm, N. 1.5 thickness for confocal microscopy.
15. Sealing of slides: Addition-curing duplicating silicone (*see* **Note 4**).
16. 1000 μL adjustable-volume micropipette.
17. Disposable pipette tips (1000 μL and 200 μL volumes).
18. High-precision micro tweezers.
19. Parafilm M.
20. Aluminum foil.

2.3 Starch-mPS-PI Staining Method for the Analysis of Arabidopsis *Seed Cellular Structures and Starch Accumulation Spatial Patterns*

1. Use the same materials described in Subheading 2.2, except for 3 and 4.

2.4 SR2200 Staining Method for the Rapid Analysis of Arabidopsis *Seed Cellular Structures (See Note 5)*

1. Fixative solution: GreenFix (Glyoxal-based formalin fixative substitute) [19] (*see* **Note 1**).
2. NaOH/SDS buffer: 0.2 N NaOH and 1% (w/v) SDS (*see* **Note 2**).
3. Bleaching solution: 1.25% (v/v) sodium hypochlorite (NaClO) in water.

4. SR2200 solution: 1% (v/v) SR2200 commercial solution (Renaissance Chemicals) in water. Store at room temperature (RT) protected from the light.
5. Antifadent mounting buffer: use an antifadent mounting buffer, preferentially prepared in a glycerol/PBS (phosphate-buffered saline) solution. We use the Citifluor™ AF1 solution in our preparations (*see* **Note 6**).
6. The same materials that were described above in points from 10 to 20 of Subheading 2.2.

2.5 Confocal Laser Scanning Microscopy

1. Leica SP5 (Leica TCS SP5) or Leica SP8 (Leica TCS SP8) laser scanning confocal microscopes, or equivalent.
2. Leica Plan Apo 20×/0.70 CS immersion correction (oil immersion) and Leica Plan Apo 40×/1.25–0.75 oil objective lenses.
3. Lasers (405 nm UV laser, 488 nm, and 561 nm lasers).
4. Leica PMT and transmitted light detectors.
5. ImageJ software (NIH, Bethesda, MD, http://rsb.info.nih.gov/ij/) with the plugin Volume Viewer (*see* **Note 7**).

3 Methods

3.1 Silique Harvesting

3.1.1 Flower Labeling

1. Use from three- to four-week-old plants to labeling flowers.
2. Select flowers at the beginning of flowering (0 days after Flowering, DAF) with the aid of a magnifying glass. Zero DAF flowers do not show white petals, as only the tip of the pistil is emerging out of closed sepals. Zero DAF flowers bear ovules at the stage 3-V of ovule development [20].
3. Label 0 DAF flowers by gently tagging the pedicel with a piece of humidified soft thread or micropore tape (Fig. 1a).

3.1.2 Silique Dissection

1. Prepare a microscope slide with double-sided tape attached to it. The tape will gently hold the silique during its dissection under the stereomicroscope (Fig. 1b–e).
2. Collect individual siliques from 0 to 8 DAF (*see* **Note 8**) using microtweezers to hold them from the pedicel, and laboratory dissecting scissors to excise them.
3. Place the silique vertically on the double-sided tape using microtweezers, positioning the stigma facing upward and the pedicel downward (Fig. 1b).
4. Thoroughly open the silique by sliding a syringe needle tip along both sides of the dehiscence zone (margins between

Fig. 1 Initial procedures and some materials required to carry out the mPS-PI and SR2200 techniques. (**a**) *Arabidopsis* flower labelling at 0 DAF with a soft thread. (**b–e**) Silique dissection steps on a double-sided tape to stain ovules and seeds by both methods. The cyan dotted lines show the dehiscence zones to be cut for silique dissection. (**f**) Tissue culture plates containing the incubation baskets used for all the treatments and washings in both staining techniques. (**g**) Sample mounting and sealing of microscope slides for their further visualization on a confocal microscope (*see* Methods)

replum and valves) while gently holding the silique on the slide with microtweezers (Fig. 1b).

5. Separate the dissected valves with a syringe needle and microtweezers (Fig. 1c, d).
6. Harvest the silique replums and septums carrying the seeds (Fig. 1e) by using microtweezers and collect them in culture well plates with flat bottom, which contain medium incubation baskets filled with 800 μL (for a 24-well plate format) of fixative solution (Fig. 1f) (*see* **Note 1**).

3.2 mPS-PI Staining (See Note 9)

3.2.1 Fixation and Tannin Precipitation

1. Seal the plate with parafilm and keep the seeds in the fixative solution for at least 1 h at room temperature (RT). Then, perform three washes in water of 5 min each, moving the incubation baskets through the different solutions with microtweezers. Optionally, samples can be kept at 4 °C overnight.

2. Incubate the samples in 800 μL of NaOH/SDS buffer for 2 h (ovules and seeds up to the heart embryo stage) or for 3 h (older seeds) at 37 °C.
3. Wash the samples three times in water (5 min per wash) (*see* **Note 10**).

3.2.2 Starch Digestion

1. Incubate the samples in 760 μL of starch digestion buffer and 40 μL of starch digestion solution for 2 h at 37 °C.
2. Wash the samples three times in water (5 min per wash) (*see* **Note 11**).

3.2.3 Tanning Bleaching

1. Incubate the samples in 800 μL of bleaching solution freshly prepared from 10 to 30 min until the tannin brownish color becomes white.
2. Wash the samples three times in water (5 min per wash) (*see* **Note 12**).

3.2.4 Aldehyde Oxidation

1. Treat the bleached samples with 800 μL of 1% periodic acid solution for 20 min with soft shaking (~100 rpm).
2. Wash the samples three times in water (5 min per wash) (*see* **Note 11**).

3.2.5 mPS-PI Staining

1. Incubate the samples in 760 μL of PS solution (*see* **Note 13**) for 10 min at RT.
2. Add 40 μL of the PI stock solution for a final volume of 800 μL and incubate samples from 2 to 4 h at RT with soft shaking (~100 rpm) (*see* **Note 14**).
3. Wash the samples three times in water (5 min per wash) and proceed to mount the slides (*see* Subheading 3.5).

3.3 Starch-mPS-PI Staining

Follow all the steps described above for mPS-PI staining, but skip the starch digestion step (*see* Subheading 3.2.2).

3.4 SR2200 Staining (See Note 15)

1. Follow the steps in Subheadings 3.2.1 and 3.2.3 described above.
2. Incubate the samples overnight with 800 μL of SR2200 solution at 4 °C (*see* **Note 16**).
3. Wash the samples three times in water (5 min per wash) and proceed to mount the slides (*see* Subheading 3.5).

3.5 Slide Mounting

1. Prepare one microscope slide with two fragments of double-sided tape positioned on either short side of the slide to attach later the coverslip (Fig. 1g).
2. For the mPS-PI techniques, add from 100 to 150 μL of chloral hydrate mounting buffer (work under a chemical fume hood).

3. For the SR2200-stained samples, add 100–150 μL of antifadent mounting buffer.
4. Locate the samples on the slide between the tape fragments with microtweezers, avoiding to place them close to the slide borders.
5. Place gently the coverslip on the slide with the aid of microtweezers starting from one tape side.
6. Mix the silicone A (catalyst) + B (base) components of twinsil® speed 22 (Picodent), a commercial mixture of polyorganosiloxanes, fillers, and additives, in a 1:1 ratio, and carefully seal the mounted slides using a 200 μL pipette tip (Fig. 1g, bottom image) (*see* **Notes 4** and **17**).
7. After 5 min (time for silicone polymerization), visualize the samples by CLSM or store the slide at 4 °C for further observations.

3.6 CLSM Imaging

3.6.1 General Microscope Settings

The settings described below allow to obtain 3D z-stack CLSM images of seeds stained with the mPS-PI and SR2200 methods.

1. Image the prepared slides with a CLSM. We have used either a Leica SP5 or a Leica SP8 confocal microscope.
2. Select the 20× objective to localize the samples or to acquire images of bigger seeds and change to the 40× objective to acquire images of ovules, early developing seeds, or specific seed tissues.
3. Set the gain and offset of each channel to balance the signal intensity and background noise level (*see* **Note 18**).
4. Display the bright field, fluorescent, and merged channels on the screen before starting the first image acquisition, and use the rainbow color option to set the exposition in the first two channels without overexposing pixels.
5. Adjust the speed of the acquisition to 100 Hz to acquire high-quality images or 400 Hz to acquire z-stack sequences (*see* **Note 19**).
6. Select the bidirectional option to scan each image in both directions, reducing acquisition time and photobleaching.
7. Adjust the line average from 4 to 8 for format sizes of up to 1024 × 1024 to compensate the signal-to-noise ratio by averaging repeat scans together (*see* **Note 20**).
8. Before running a z-stack acquisition, set the beginning and the end of the stack moving along the *z* axis.
9. Activate and set the z-compensation for the laser power and gain when acquiring thick z-stacks showing variable levels of signal through the *z* axis.

3.6.2 Specific Settings for the Acquisition of mPS-PI Images

Excite mPS-PI-stained samples at 488 nm and 561 nm, and collect the emission between 530 nm and 730 nm.

3.6.3 Specific Settings for the Acquisition of SR2200 Images

Excite the SR2200-stained samples at 405 nm (UV), and collect the emission between 415 nm and 476 nm.

3.7 Image Processing

Process the images using the ImageJ software. The volume viewer plug-in should be installed to perform the z-stack analysis.

1. Open the z-stack series with volume viewer. Rotate the 3D image along the *x*, *y*, and *z* axes in order to obtain the desired seed orientation. Then, move along the "distance" setting to obtain the desired 2D optical section (Fig. 2c, f and 3c).
2. Optionally, the volume 4 mode of volume viewer can be used to acquire three-dimensional images (Fig. 3d–f).

Figures 2 and 3 show examples of confocal images of Col-0 seeds at 3 DAF obtained with the mPS-PI and SR2200 staining methods.

4 Notes

1. Use the fixative solution (*see* Subheading 3.2.1) in a fume chemical hood. It is a nontoxic but highly volatile compound.
2. The SDS in the NaOH/SDS buffer can precipitate at low temperature. To redissolve it, incubate the solution at 37 °C as long as necessary.
3. We recommend using a fresh α-amylase solution (do not store it for more than a week at 4 °C) because it is a sensitive enzyme. If the enzyme loses its activity, residual starch signal might be observed when performing the mPS-PI staining technique.
4. We recommend using the twinsil® speed 22 (Picodent) as previously reported [21].
5. The SR2200 solution can be replaced with a 10 μg/mL Calcofluor White M2R water solution. Calcofluor-stained samples require CLSM acquisition parameters [17] similar to those described for SR2200-stained samples [16]. The only difference is the emission collected between 415 and 448 nm [17].
6. SR2200-stained samples can be mounted in water. Nevertheless, we suggest mounting them in an antifadent solution, preferably containing an antibleaching agent, such as the Citifluor™ AF1.

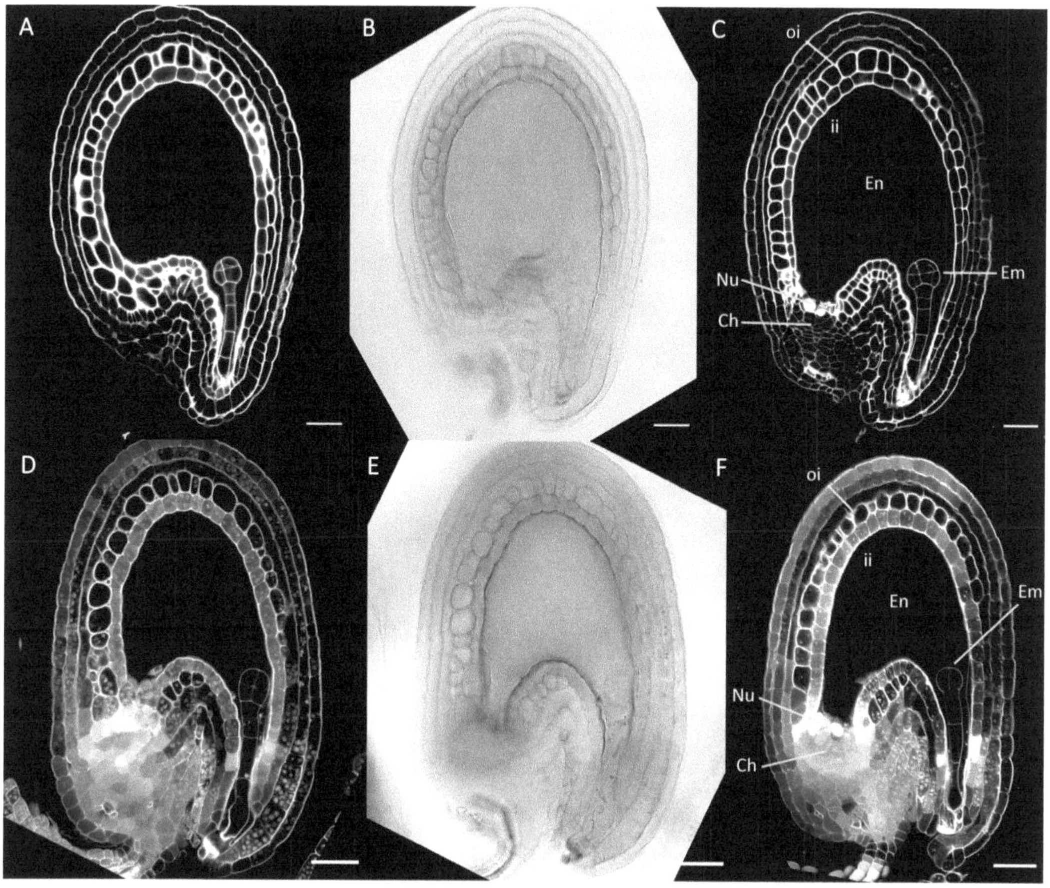

Fig. 2 Col-0 seeds at 3 DAF stained by the mPS-PI techniques. (**a–c**) mPS-PI stained seeds showing highlighted cell walls of different cell types at the globular stage of growth. (**a**) PI channel, (**b**) Bright field and (**c**) Central sagittal section of Col-0 *Arabidopsis* seeds. (**d–f**) Starch and cell walls were stained in seeds at the globular stage by the mPS-PI technique without α-amylase (*see* Methods). (**d**) PI channel, (**e**) Bright field and (**f**) Central sagittal section of Col-0 seeds. *Ch* Chalaza, *Nu* Nucellus, *oi* Outer integuments and *ii* Inner integuments of the seed coat, *En* Endosperm and *Em* Embryo. Scale bars: 30 μM

7. Alternatively, confocal images can be processed with more advanced image analysis software, such as Imaris or LAS X.
8. The SR2200-staining protocol works only with *Arabidopsis* seeds at early developmental stages (until 6 DAF, heart embryo stage). Older seeds must be imaged using the mPS-PI staining technique [11].
9. The mPS-PI and SR2200 staining protocols described here are compatible with GUS but not GFP imaging analyses. Alternative GFP-compatible techniques, such as ClearSee [22], have been developed, but they are longer and more labor intensive.
10. We do not recommend keeping seeds at 4 °C for longer than overnight after the three washes following the NaOH/SDS

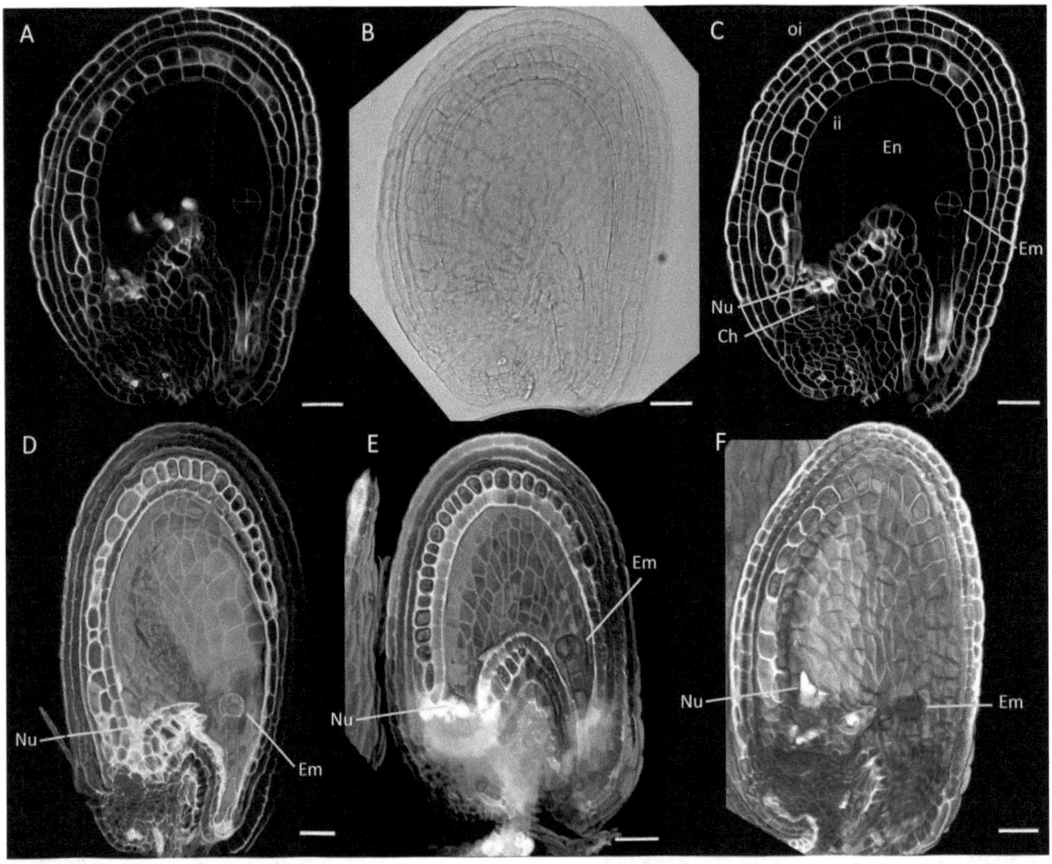

Fig. 3 SR2200 stained seeds at 3 DAF and 3D-images of mPS-PI and SR2200 staining. (**a–c**) SR2200 stained seeds showing defined cell walls of different cellular groups at the globular stage of development. (**a**) PI channel, (**b**) Bright field and (**c**) Central sagittal section of Col-0 *Arabidopsis* seeds. (**d–f**) 3D-images of central sagittal sections of mPS-PI and SR2200 stained seeds displaying contrasting cells. (**d**) mPS-PI, (**e**) mPS-PI without α-amylase and (**f**) SR2200 stained seeds at 3 DAF. *Ch* Chalaza, *Nu* Nucellus, *oi* Outer integuments and *ii* Inner integuments of the seed coat, *En* Endosperm and *Em* Embryo. Scale bars: 30 μM

treatment, as SDS residues might precipitate and interfere with subsequent staining steps.

11. After this step, samples can be kept overnight at 4 °C, if necessary.
12. We suggest washing the samples five times to remove any bleach residue when cell wall staining is combined with other histological dyes, such as Auramine O [23].
13. Store the pseudo-Schiff solution for up to 4 months at RT. Longer storage periods might affect the activity of the solution and require longer incubation time with PI to stain the samples, affecting seed integrity.

14. For mPS-PI staining of seeds older than 4 DAF, we recommend a 3–4 h incubation time in the PS-PI solution to ensure the penetration of PI in the seed.
15. SR2200- or Calcofluor White M2R staining can be combined with other histological dyes [18], such as Auramine O [17, 23].
16. For SR2200 staining of seeds older than 3 DAF, we recommend extending the dye incubation to 48 h at 4 °C.
17. Slides of SR2200-stained samples can be sealed using nail polish. Nevertheless, this sealing technique does not prevent slides dehydration for extended periods (more than 15 days). We do not recommend using nail polish for sealing slides with mPS-PI-stained samples, as the solution can easily penetrate between slide and coverslip and mix with chloral hydrate, thus affecting the seed coat integrity.
18. When the fluorescence signal is weak in some areas of the sample, we recommend increasing the laser power instead of the gain. Then, gain and offset can be readjusted for each channel.
19. If using the Leica SP8 confocal microscope, speed can be increased up to 600 Hz without altering the zoom factor. We advise to use this speed when capturing thick z-stacks to reduce photobleaching.
20. When consecutive scans are averaged, the pixel exposure time increases, thus improving image quality. For format sizes bigger than 1024 × 1024, it is necessary to increase the number of line averages (e.g., up to 16 for a 2048 × 2048 format).

Acknowledgments

This work has benefited from the support of IJPB's Plant Observatory technological platforms and was founded by the PERISEED ANR (ANR-18-CE20-0009) and Labex Saclay Plant Sciences-SPS (ANR-10-LABX-0040-SPS) grants.
The IJPB benefits from the support of Saclay Plant Sciences-SPS (ANR-17-EUR-0007).

References

1. O'Brien IEW, Reutelingsperger CPM, Holdaway KM (1997) Annexin-V and TUNEL use in monitoring the progression of apoptosis in plants. Cytometry 29:28–33
2. O'Brien IEW, Murray BG, Baguley BC et al (1998) Major changes in chromatin condensation suggest the presence of an apoptotic pathway in plant cells. Exp Cell Res 241: 46–54
3. Riccardi C, Nicoletti I (2006) Analysis of apoptosis by propidium iodide staining and flow cytometry. Nat Protoc 1:1458–1461

4. Zhu H (2012) Propidium iodide staining of cells for FACS analysis. Bio Protocol 2. https://doi.org/10.21769/BioProtoc.195
5. Darzynkiewicz Z, Huang X, Zhao H (2017) Analysis of cellular DNA content by flow cytometry. Curr Protoc Immunol 2017: 5.7.1–5.7.20
6. Jones K, Kim DW, Park JS et al (2016) Live-cell fluorescence imaging to investigate the dynamics of plant cell death during infection by the rice blast fungus *Magnaporthe oryzae*. BMC Plant Biol 16:1–8
7. Truernit E, Bauby H, Dubreucq B et al (2008) High-resolution whole-mount imaging of three-dimensional tissue organization and gene expression enables the study of phloem development and structure in Arabidopsis. Plant Cell 20:1494–1503
8. Rae AE, Wei X, Flores-Rodriguez N et al (2021) Super-resolution fluorescence imaging of arabidopsis thaliana transfer cell wall ingrowths using pseudo-schiff labelling adapted for the use of different dyes. Plant Cell Physiol 61:1775–1787
9. Flütsch S, Distefano L, Santelia D (2018) Quantification of starch in guard cells of *Arabidopsis thaliana*. Bio Protoc 8:e2920
10. Lu J, Le HR, Gómez-Páez DM et al (2021) The nucellus: between cell elimination and sugar transport. Plant Physiol 185:478–490
11. Xu W, Fiume E, Coen O et al (2016) Endosperm and nucellus develop antagonistically in Arabidopsis seeds. Plant Cell 28:1343–1360
12. Coen O, Lu J, Xu W et al (2020) A TRANSPARENT TESTA transcriptional module regulates endothelium polarity. Front Plant Sci 10: 1–14
13. Haseloff J (2003) BioImaging old botanical techniques for new microscopes. Biotechniques 34:1174–1182
14. Moreno N, Bougourd S, Haseloff J et al (2006) Imaging plant cells. Handbook of biological confocal microscopy, vol 166, 3rd edn, pp 769–787
15. Truernit E, Siemering KR, Hodge S et al (2006) A map of KNAT gene expression in the Arabidopsis root. Plant Mol Biol 60:1–20
16. Musielak TJ, Schenkel L, Kolb M et al (2015) A simple and versatile cell wall staining protocol to study plant reproduction. Plant Reprod 28:161–169
17. Buda GJ, Isaacson T, Matas AJ et al (2009) Three-dimensional imaging of plant cuticle architecture using confocal scanning laser microscopy. Plant J 60:378–385
18. Tofanelli R, Vijayan A, Scholz S et al (2019) Protocol for rapid clearing and staining of fixed Arabidopsis ovules for improved imaging by confocal laser scanning microscopy. Plant Methods 15:1–13
19. Benerini Gatta L, Cadei M, Balzarini P et al (2012) Application of alternative fixatives to formalin in diagnostic pathology. Eur J Histochem 56(2):e12
20. Schneitz K, Hülskamp M, Pruitt RE (1995) Wild-type ovule development in *Arabidopsis thaliana*: a light microscope study of cleared whole-mount tissue. Plant J 7:731–749
21. Provost A, Rousset C, Bourdon L et al (2019) Innovative particle standards and long-lived imaging for 2D and 3D dSTORM. Sci Rep 9: 1–13
22. Ursache R, Andersen TG, Marhavý P et al (2018) A protocol for combining fluorescent proteins with histological stains for diverse cell wall components. Plant J 93:399–412
23. Coen O, Lu J, Xu W et al (2019) Deposition of a cutin apoplastic barrier separating seed maternal and zygotic tissues. BMC Plant Biol 19:1–20

Part IV

Genetic Analyses and Gene Manipulation

Chapter 10

Map-Based Cloning of the Causal Gene for a Seed Dormancy Quantitative Trait Locus in Barley

Shingo Nakamura and Kazuhiro Sato

Abstract

Seed dormancy is an important agronomic trait in cereal crops. Throughout the domestication of cereals, seed dormancy has been reduced to obtain uniform germination. However, grain crops must retain moderate levels of seed dormancy to prevent problems such as preharvest sprouting in wheat (*Triticum aestivum*) and barley (*Hordeum vulgare*). To produce modern cultivars with the appropriate seed dormancy levels, it is important to identify the genes responsible for seed dormancy. With recent advances in sequencing technology, several causal genes for seed dormancy quantitative trait loci (QTLs) have been identified in barley and wheat. Here, we present a method to identify causal genes for seed dormancy QTLs in barley, a method that is also applicable to other cereals.

Key words Seed dormancy, Preharvest sprouting, Natural variation, Quantitative trait locus, Barley

1 Introduction

Seed dormancy controls the timing of germination, which is important for plant survival and adaptation to the surrounding environment. During domestication, farmers selected cereal varieties with lower levels of seed dormancy to obtain uniform germination at sowing. However, low seed dormancy can cause problems such as preharvest sprouting (PHS), in which seeds germinate on the plant before harvest, reducing grain yield and quality. PHS is a serious problem in cereal crops, especially in wheat (*Triticum aestivum*) and barley (*Hordeum vulgare*) produced in regions where rainy or humid conditions tend to occur during the harvest season [1]. Conferring the appropriate levels of seed dormancy to allow uniform germination and prevent PHS remains an important objective in wheat and barley breeding programs in such regions.

Seed dormancy is affected by numerous quantitative trait loci (QTLs), and to promote wheat and barley breeding programs, it is important to identify the QTLs and associated causal genes that

Naoto Kawakami and Kazuhiro Sato (eds.), *Seed Dormancy: Methods and Protocols*, Methods in Molecular Biology, vol. 2830, https://doi.org/10.1007/978-1-0716-3965-8_10,

control seed dormancy [2]. Sequenced reference genomes provide physical maps covering the entire genome, facilitating efforts to identify QTLs of interest, and researchers have published the large and complex genomic sequences of wheat and barley, as well as their first-generation pan-genome assemblies [3, 4]. In addition, powerful new technologies, such as high-throughput short- and long-read sequencing, have enabled us to easily access millions of polymorphisms covering entire genomes, and to efficiently genotype thousands of plants with numerous insertion/deletion (Indel) and single-nucleotide polymorphism (SNP) markers [5]. These technological advances have greatly facilitated the identification of the causal genes for QTLs that govern important agronomic traits in barley and wheat as well as in other cereals.

Barley has a wide range of variation in seed dormancy, from the extremely strong dormancy of the wild progenitor of cultivated barley (*H. vulgare* ssp. *spontaneum*) to the reduced dormancy in malting barley [6]. Many research programs have tried to identify the causal genes for seed dormancy QTLs in barley and wheat since QTL analysis was first reported in the early 1990s, using the restriction fragment length polymorphism (RFLP) markers developed in the 1980s. However, it took more than 20 years to finally identify several causal genes in major seed dormancy QTLs in barley and wheat [2, 7–11]. The results are being used in breeding programs to develop PHS-tolerant barley and wheat varieties. Despite much recent progress on these major seed dormancy QTLs, a number of seed dormancy QTLs remain to be identified in barley and wheat as well as other cereals.

Here, we present a method to identify causal genes for seed dormancy QTLs in barley by a map-based cloning method using progeny test by following procedures. We grow a mapping population derived from the cross between parents with different alleles on target QTL. We collect seed and DNA samples for dormancy evaluation and genotyping. A linkage map is constructed on the mapping population with DNA markers. We score the level of seed dormancy by germination and identify the QTL on the linkage map (Figs. 1 and 2). The seed dormancy QTL is fine mapped by using lines with introgressed segments that cover the QTL. A candidate gene is evaluated by transformation and/or mutation experiments.

2 Materials

2.1 Plant Materials

1. Obtain seeds of parental varieties. Seeds can be requested from a gene bank if needed. In addition, prepare recombinant inbred lines (RILs) (F_6 or later generations of single-seed descent) and near isogenic lines (NILs) (BC_3 or BC_4F_6 or later self-pollinated generations) derived from the parental varieties (*see* **Note 1**). If the RIL and NIL populations are not available,

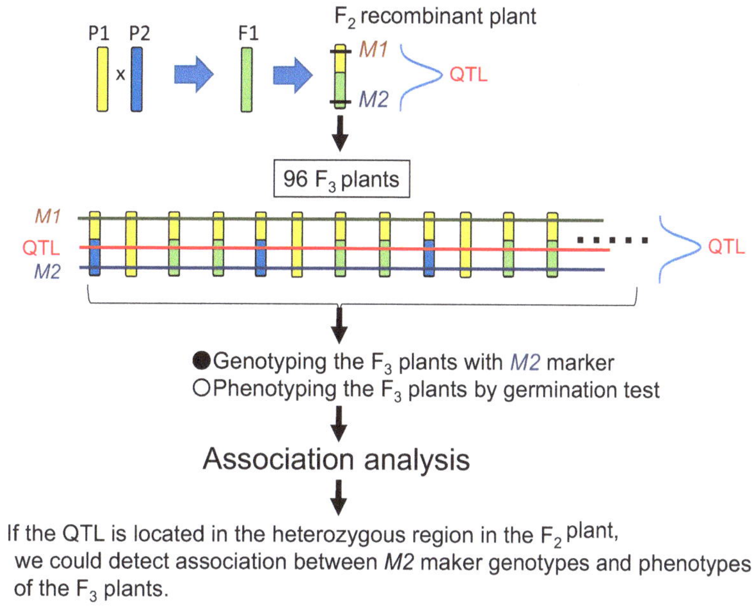

Fig. 1 A summary of the association analysis method for fine mapping a seed dormancy QTL. Genotype of a chromosome that has a targeted seed dormancy QTL. Yellow: Parent 1 variety homozygous region, Blue: Parent 2 variety homozygous region, Green: heterozygous chromosome region. The red solid line indicates the location of the QTL. *M1* and *M2* are DNA markers located near the QTL

make them by crossing the parental varieties; this can take several years. Practical sizes for the RIL and NIL populations are 96 or 192 because 96-well plates are typically used for genotyping experiments.

2.2 Growing Barley in Fields(See Note 2)

1. Important aspects of field management to consider include fertilizers (e.g., advanced fertilizer 6–9-6, granular-fused magnesium phosphate, magnesium lime), insecticides (e.g., Diazinon or Sumithione (fenitrothion) for controlling aphids), broadleaf herbicides for cereals (e.g., Galace (difulfenican)), fungicide (e.g., Silvacur flowable) for controlling fusarium), bird nets (mesh size 2 × 2 cm) for preventing birds, 50-cm fiberglass poles for numbering plants, and labels for displaying line numbers and names.
2. For breaking seed dormancy before sowing, use 1% (w/w) hydrogen peroxide (H_2O_2).

   Dilute 30% (w/w) H_2O_2 with distilled water to make 1% (w/w) H_2O_2.

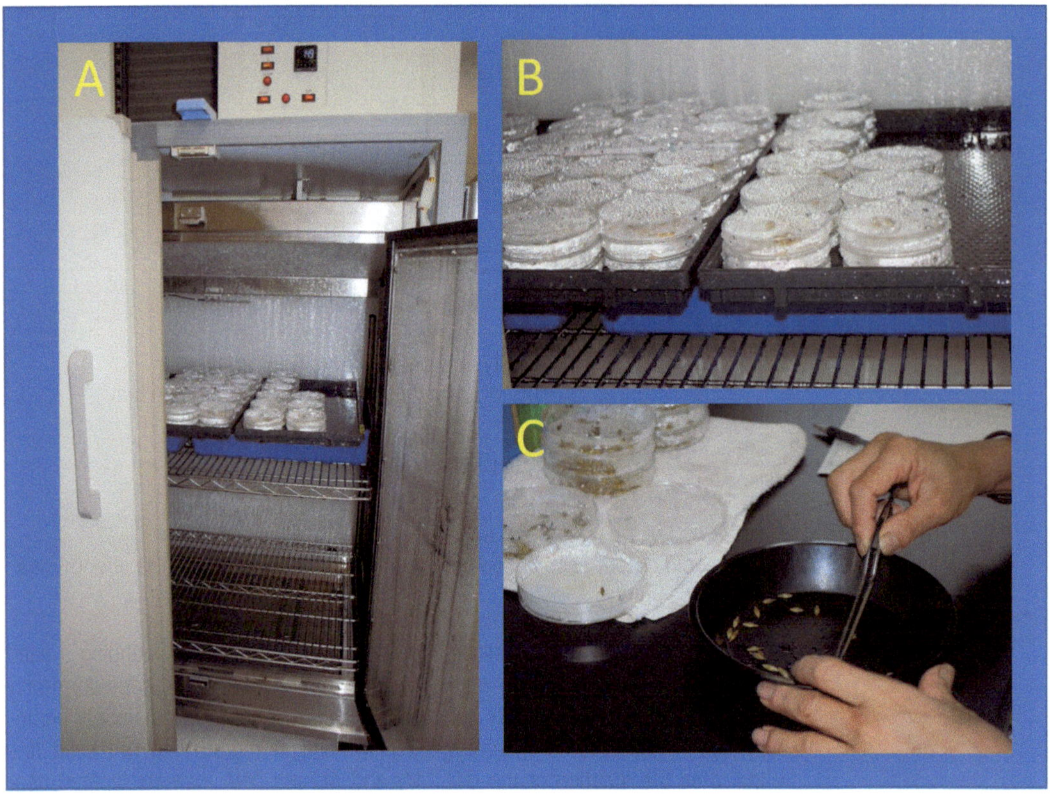

Fig. 2 Germination test. (**a**) and (**b**) Seeds set in Petri dishes were being incubated at 15 °C in an incubator with controlled humidity to keep 100% humidity in the growth chamber so that no seals or vinyl bags to wrap the Petri dishes are needed to prevent evaporation of water from the Petri dishes (**c**) After 7 days incubation, germinated seeds were counted using tweezers

2.3 The Simple Genomic DNA Extraction Method (See Note 3)

1. 96-well PCR plates (0.2 mL wells).
2. Electronic autopipettes such as Picus.
3. Shaker beads such as TissueLyser.
4. Collection microtubes.
5. Glass or zirconia beads (e.g., 3-mm diameter).
6. Genomic DNA extraction buffer: 200 mM Tris–HCl pH 7.5, 250 mM NaCl, 0.5% SDS, 25 mM EDTA. For preparing 1000 mL of extraction buffer, mix 200 mL of 1 M Tris–HCl pH 7.5, 50 mL of 5 M NaCl, 50 mL of 10% SDS, 50 mL of 0.5 M EDTA, and 650 mL of distilled deionized water.

2.4 Genotyping

1. This depends on the technique used. For example, for a classical technique, you will need cleaved amplified polymorphic sequence (CAPS) markers, oligonucleotides, a PCR machine, and an agarose gel system. Alternatively, you could use a high-throughput genotyping system such as high-density 50 K SNP

array with the Illumina Infinium whole-genome typing array [12].

2.5 Construction of a Linkage Map

1. Obtain genotype data for the RILs.
2. Software such as a JoinMap can be used for construction of a linkage map (*see* **Note 4**).

2.6 Germination Test (See Note 5)

1. Filter paper (e.g., 85-mm diameter, No. 2).
2. Petri dishes (e.g., 90-mm diameter (φ) × 15-mm height).
3. Distilled water (e.g., 4 mL per Petri dish (*see* **Note 6**)).
4. Fungicide (e.g., 0.025% iminoctadine triacetate).
5. An incubator. For convenience, choose an incubator with controlled humidity so you do not have to seal the Petri dishes.

2.7 QTL Analysis

1. Barley seeds of parental varieties (*see* **Note** 7). Mapping populations (RILs) derived from the parental varieties (*see* **Note 1**).
2. Equipment to cross plants (fine-point precision tweezers, scissors, paper envelope or bag, marking pen, a bottle of water).
3. Equipment to grow plants in the field (*see* **Note 8**).
4. The simple genomic DNA extraction method.
5. Equipment and materials for genotyping (e.g., PCR systems, agarose gel electrophoresis) or a high-throughput genotyping system (e.g., array genotyping system).
6. Software for constructing the linkage map and QTL analysis (e.g., JoinMap, MapQTL: https://www.kyazma.nl/index.php/About/).

2.8 Fine Mapping QTLs

1. NILs derived from the parental varieties (*see* **Note 1**). F_2 plants derived from a cross between the NILs and their recurrent parental line. F_3 seeds derived from the F_2 plants.
2. Equipment to cross plants (*see* above).
3. Equipment to grow plants in fields (*see* above).
4. Genetic markers (e.g., CAPS markers, Kompetitive allele-specific PCR (KASP) markers [13]).

3 Methods

3.1 Growing Barley in Fields

1. If seeds are dormant, break seed dormancy by treating the seeds with 1% (w/w) H_2O_2 overnight at room temperature (*see* **Note 9**).
2. Prepare the field for sowing (e.g., plowing, applying fertilizers).

3. Sow winter barley seeds in autumn (e.g., sow 20 barley seeds per line at 20-cm intervals in a 4-m line. The lines are 70 cm apart).
4. Set a net on the field to prevent birds and other animals from eating the seeds or seedlings.
5. Tread growing barley several times every 30 days starting a month after sowing [14].
6. Spray insecticides in the spring to remove aphids if necessary.
7. Spray fungicide at the appropriate time before flowering to prevent *Fusarium* infection.
8. Set a net before flowering to prevent birds from eating the developing seeds.
9. When seeds reach physiological maturity, harvest spikes by hand with scissors. Physiological maturity occurs when the color of the peduncle turns yellow or the awn becomes dry (starting from the tip), indicating that the spike's abscission layer has fully formed, and the seeds are no longer receiving water and nutrients from the plant (*see* **Note 10**).
10. Collect the harvested spikes into paper envelopes or net bags and dry them at 30 °C for 1 week under constant airflow in a harvest oven.
11. Store the dried spikes at −30 °C to maintain their dormancy as long as possible (*see* **Note 11**).
12. Remove the spikes from the freezer, gently hand-thresh the spikes to limit damage to the seeds (*see* **Note 12**), and put seeds from each spike in an envelope (paper bag) labeled with harvest time and line name. Store the seeds in the envelopes in a freezer at −30 °C again until all seeds become ready for the germination test.

3.2 Simple Genomic DNA Extraction Method

1. Remove a 5 × 5 mm section of a young leaf from each plant and place it in a 1.5 mL microfuge tube or collection microtube. Add 3–5 glasses or 1 zirconia bead and leave the tubes open at least overnight at 50 °C to completely dry the leaves.
2. Crush the dried leaf tissue with the glass or zirconia beads using a TissueLyser for 1 min with a frequency of 25–30 per second. Reverse the rack orientation and repeat (*see* **Note 13**).
3. Add 300 μL of the DNA extraction buffer to each tube, and centrifuge the tubes for 10 min at 13,000 rpm or 15 min at 4800 rpm.
4. Add 100 μL of isopropanol to each well of a 96-well plate. Dispense the 100 μL of supernatant from each sample after the centrifugation into individual wells of the 96-well plate and mix it using the same pipette.

5. Centrifuge the 96-well plate at 4800 rpm for 15 min. Discard the supernatant and invert the plate on filter paper until all liquid has dried. Add 100 μL TE to the DNA in the wells and allow to sit for at least an hour at room temperature. Then seal the 96-well plate and store it at −30 °C. Alternatively, you can store the plate at 4 °C for up to several weeks.

3.3 Genotyping

1. Extract genomic DNA from leaves of the RILs and NILs grown in the field using the simple extraction method described above or a commercially available kit such as DNeasy Plant Mini Kit. Store the extracted DNA at −30 °C until use.
2. Genotype the RIL and NIL populations using a high-throughput genotyping system such as 1536 mapped SNPs of high-density 50 K SNP array with the Illumina Infinium whole-genome typing array [12]. BOPA1 can determine the genotypes of genomic sequences of all chromosomes of the populations with a sufficient level of marker density (*see* **Note 14**).

   If needed, other types of DNA markers, such as CAPS and KASP markers, can also be used.

3.4 Construction of a Linkage Map

1. Prepare the obtained genotyping data in an appropriate format according to the instruction manual of the software program.
2. Input the data into the chosen software program and use the program to produce linkage maps.

3.5 Germination Test

Conduct a germination test using at least three spikes per line as biological replicates.

1. Set two layers of filter paper in Petri dishes and wet the paper with 4 mL of distilled water (*see* **Note 5**) containing 0.025% iminoctadine triacetate as a fungicide.
3. Add the hand-threshed seeds from a spike to the Petri dish.
4. Incubate the dishes at 15 °C for 1 week in darkness (Fig. 2a, b, *see* also **Note 15**).
5. After incubation, count the number of germinated and non-germinated seeds (Fig. 2c), and calculate the germination percentage of each spike (*see* **Note 16**).

3.6 QTL Analysis

1. To calculate QTL data (logarithm of odds (LOD) score, etc.), process the linkage-map data and germination data from the RIL population using appropriate software programs such as MapQTL (https://www.kyazma.nl/index.php/MapQTL/: *see* also **Note 4**).
2. To test the reproducibility of the QTL, conduct the QTL analyses with samples harvested over at least three different environments (years or locations). If the QTLs are

reproducible and worth cloning, proceed further to fine mapping of the QTL.

3.7 Fine Mapping the QTL

1. Select NILs that contain introgressed genomic segments that cover the QTL (*see* **Note 17**).
2. Cross the selected NIL with its recurrent parent to produce F_1 seeds.
3. Sow the F_1 seeds in the field, grow F_1 plants, and obtain F_2 seeds (*see* **Note 18**).
4. Sow several thousand F_2 seeds in the field and grow individual F_2 plants in segregating population.
5. Harvest young leaves according to the simple genomic DNA extraction method outlined above.
6. Extract genomic DNA from each F_2 plant as outlined above.
7. Check the genotypes of the F_2 plants using two markers that flank the QTL to find plants that have a recombination in the QTL region (Fig. 1).
8. Categorize the genotypes of the F_2 plants into four recombination types depending on the genotypes of the flanking markers (shown as *M1* and *M2* in Fig. 1): I, *M1* homozygous for the genotype of parent P1 and *M2* heterozygous; II, *M1* homozygous for parent P2 and *M2* heterozygous; III, *M1* heterozygous and *M2* homozygous for parent P1; IV, *M1* heterozygous for parent P2 and *M2* homozygous. From each genotype, select several representative F_2 plants for subsequent association analysis.
9. Harvest young leaves from all F_2 plants that have a recombination in the QTL interval between the flanking markers (Fig. 1).
10. Extract genomic DNA from the selected recombinant F_2 plants for subsequent fine mapping using a commercially available kit such as DNeasy Plant Mini kit. Store the extracted genomic DNA at −30 °C.
11. Harvest more than 120 F_3 seeds from each F_2 plant that has a recombination in the QTL interval (*see* **Note 18**).
12. Sow 120 F_3 seeds from the selected representative F_2 plants in the field and grow F_3 plants.
13. Select 96 F_3 plants from the sowed 120 F_3 plants derived from each representative F_2 plant. And put a glass fiber pole at the side of every five plants. The side poles are useful to confirm which number the plant is when we harvest spikes.
14. Harvest young leaves from each F_3 plant.
15. Extract genomic DNA from the leaves using the simple extraction method outlined above.

16. Genotype the F_3 plants using one of the two markers located on the heterozygous genomic region of the F_2 plants (Fig. 1). Select F_3 plants that have a homozygous genotype for the marker. According to Mendelian segregation, a 1:2:1 segregation ratio would result in approximately 24 (parent P1 homozygous type): 48 (heterozygous type): 24 (parent P2 homozygous type) plants from the total 96 F_3 plants.
17. Harvest F_4 seeds from at least three spikes from the homozygous type F_3 plants for a germination test.
18. Conduct the germination test for the F_4 seeds from each homozygous type F_3 plant.
19. Conduct association analysis by checking associations between marker genotypes and germination percentages using Welch's t test (Fig. 1) (*see* **Notes 19, 20,** and **21**). If the results of the Welch's t test are significant, this indicates that the germination percentages are associated with the marker genotypes. This means that the QTL is located on the heterozygous side of the genomic region of the F_2 plants. The target QTL region can thus be narrowed down using the data from all selected representative F_2 plants.
20. For fine mapping, construct new markers that are located in the narrowed QTL region. Using the new markers, genotype the F_2 population to find F_2 plants with recombination events in the narrowed-down region.
21. Sow the F_3 seeds from the newly identified F_2 plants and conduct association analysis again. Repeat the process until all recombinant F_2 plants are analyzed. To be able to identify a candidate gene, the QTL needs to be delimited to a very narrow region, such as several kbp around a gene region. If the region is not narrow enough, new F_2 plants that have a recombination in the narrowed-down genomic region will need to be identified through further screening of newly created F_2 plants. Continue fine mapping with the new F_2 recombinant plants until the genomic region is narrow enough. If new recombinant F_2 plants cannot be identified, consider other ways to identify the causal gene for the QTL. For example, if a promising candidate gene is identified in the region, evaluate the gene using complementation cloning or other methods like using TILLING mutants or the CRISPR/Cas9 system [15] (*see* **Note 22**).

3.8 Complementation Test for Functional Evaluation of the Identified Candidate Genes

1. Amplify a genomic fragment that covers the entire candidate gene region identified by fine mapping using a PCR system and a high-fidelity DNA polymerase (*see* **Note 23**).
2. Clone the fragment into an appropriate *Agrobacterium* transformation binary vector.

3. Sequence the cloned fragment to confirm that the sequence is correct.
4. Transfer the constructed binary vector into a backcross-derived line (*see* **Notes 24** and **25**) or transform amenable cultivars using the *Agrobacterium*-mediated transformation system [5, 16, 17].
5. Grow transformed T_0, T_1, and T_2 generation plants in growth chambers.
6. Check their dormancy levels using a germination test and analyze the relationship between the introduced candidate gene and the seed dormancy level. An indication that the inserted genomic fragment affects seed dormancy of the transformed plants is strong supportive evidence that the candidate gene is the true causal gene of the QTL (*see* **Note 26**).

4 Notes

1. Instead of RILs and NILs, doubled haploid lines (DHLs) and chromosome segment substitution lines (CSSLs) can also be used. Use of NILs and CSSLs as materials for fine mapping is important because these lines minimize the effects of other nontarget QTLs on the phenotype. In addition, consider which parental genomic segment is incorporated into which parent's background. For this, consider which QTL allele shows more evident and stable phenotypes in the NILs or CSSLs. In the case where QTL alleles show dominance, it is important to choose NILs or CSSLs that contain genomic segments containing the dominant QTL allele. For this, you first need to know which parent contains the dominant QTL allele.
2. Consider the growing conditions for the selected plant materials to maximize the QTL effects for more accurate phenotyping. If a single QTL effect is stably detected in field conditions, this may be sufficient for identifying the plants with the desired QTL allele. However, if the effect of a single QTL is too weak to identify in field conditions, the QTL may be affected by environmental factors. In this case, consider growing plant materials under controlled environmental conditions to minimize environmental effects as much as possible. A good example of this is the analysis of the wheat seed dormancy QTL *Phs1* [10]. They found that the QTL effect did not express well in field conditions but did express well in controlled greenhouse conditions [18, 19]. Thus, they carried out fine mapping of the QTL on greenhouse-grown plants and eventually identified the causal gene for the QTL as *MKK3* [10].

3. A commercially available high-throughput genomic DNA extraction kit can also be used.
4. Refer to the software program list in the reference [20].
5. Preharvest sprouting tolerance can be determined using seeds still on the spike and beforehand-threshing. A new wheat preharvest sprouting tolerance QTL was recently reported using this method [21].
6. Do not add too much water in the Petri dishes because this will inhibit germination of barley seed.
7. Information on the germplasm and genetic stocks can be found through the GrainGenes database (https://wheat.pw.usda.gov/GG3/germplasm#barley).
8. Greenhouses or incubators can also be used if they are suitable for growing barley.
9. This 1% (w/w) H_2O_2 method can be used to break dormancy of seeds of the wild progenitor of cultivated barley (*H. vulgare* ssp. *spontaneum*), which have extremely strong dormancy.
10. The seed development stage is also important. In general, seeds will germinate well after the water content decreases from around 40% to 20% or less at physiological maturity in barley. Thus, to detect seed dormancy QTLs, it is important to use physiologically matured dried seeds. Seed germination conditions (temperature, amount of water and light) are also important. For example, in wheat and barley, lower temperatures stimulate germination. If you use too low a temperature, every seed, even those with strong dormancy, would germinate. Thus, select an appropriate temperature that maximizes the difference in germination between the parental varieties chosen for QTL analysis.
11. In some cases, seed dormancy level cannot be completely maintained in long-term storage (e.g., many years) even when stored at −30 °C. Therefore, it is better to conduct the germination test within a year or so after harvest.
12. Wounding seeds stimulates germination.
13. If using a machine from a different manufacturer, follow the manufacturer's instructions.
14. Other high-throughput genotyping systems can also be used.
15. Light inhibits germination in barley [22]. Thus, seeds are incubated in darkness.
16. In general, the result of the germination test is expressed in two ways. One way is to measure the percentage of germinated seeds after a certain length of incubation time, e.g., a week. The other way is to use the germination index (GI). This index gives the maximum weight of seeds that germinate rapidly and

is calculated from the following formula: GI = (T × N_1 + (T−1) × N_2 + (T−2) × N_3 + · · · · · · · · + 2 × $N_{(T-1)}$ + 1 × N_T)/(T × total number of seeds), where T is the number of total days of the test, N_1 to N_T are the number of seeds that germinated from day 1 to final day T. The maximum index is 1.0 if all seeds germinate by day 1 while lower indices are indicative of an increasing level of seed dormancy or reduced germinability.

17. The introgressed segment size should not be too small because short fragments can cause the recombination to be suppressed in the genomic region. Thus, fragments around 20 cm in length are preferred.
18. If possible, it is better to cover the spikes with envelopes before flowering to prevent out-crossing of F_2 seeds.
19. In this situation, it is important to be able to clearly detect and evaluate the phenotypic effect derived from the targeted single QTL for successful pursuit of the causal gene of the QTL of interest or subsequently to identify functional sequence polymorphisms. Therefore, in the case of seed dormancy traits, it is critical to be able to evaluate differences in phenotypic effects from dormant and nondormant QTL alleles.
20. Linkage map construction software can also be used to evaluate the associations by LOD score value.
21. Statistical methods must be used to evaluate the genotypes of the QTL of interest. Unlike qualitative discrete traits, such as classical round/wrinkled skin in pea, it is impossible to directly determine dormant and nondormant QTL alleles from a germination test because the germination percentage can fluctuate due to complex interactions of genetic and environmental factors. Thus, to determine the genotype of the targeted QTL for fine mapping, statistical methods are used to evaluate the association of genotypes of markers located near the target QTL and the phenotypes of dozens of progeny plants as described in the Methods section.
22. Once the gene region has been finely mapped to within several kb, it is important to examine gene function to confirm the candidate gene. This is usually done via a genetic complementation test or by evaluating the phenotypes of plants that have mutations in the identified gene. Finding or generating at least three independent mutants of the candidate gene will increase confidence in the results. The TILLING method for screening barley mutants is described in Jost et al. (2019).
23. A DNA polymerase for high-fidelity and long PCR, such as Prime STAR GXL DNA polymerase, can be used.

24. In barley, only a limited number of varieties are highly amenable to transformation, such as cultivars "Golden Promise" or "Igri." However, in the majority of cases the variety containing the (partial) loss-of-function (recessive) allele of the QTL is recalcitrant to transformation. Thus, in most cases, the transformation cannot be conducted directly on the variety carrying the recessive allele of the QTL. One way to circumvent this is to develop, in advance, backcross-derived lines in the background of an amenable variety that contains introgressed chromosome segments carrying the recessive allele. If the developed backcross-derived lines are amenable to transformation and can show the QTL effects when transformed by the functional (dominant) allele of the QTL, they can be used for complementation cloning. The study by Sato et al. (2016) is an example of this strategy being used successfully in cloning the seed dormancy QTL *Qsd1* [11].
25. Barley genomic regions responsible for the amenability to transformation have been reported [5, 23, 24], and can be used to select backcross-derived lines.
26. If a recessive allele of the QTL shows complete loss of function, such as a nonsense mutation, genome editing technology can be used to knock down the candidate gene in transformation-amenable cultivars to prove gene function. However, even in that case, differences in the genetic backgrounds of different varieties used for QTL analysis and transformation experiments may complicate interpretation of the results because the different backgrounds may mask the original QTL effect. In addition, in the case where another gene that affects seed dormancy is located near the causal QTL gene of interest, gene knockout could give misleading results, as knockout of either gene could affect seed dormancy. Thus, it is important to home in on the location of the QTL, defining a very narrow region, when using the complementation test strategy to evaluate the function of the candidate QTL gene.

References

1. Nonogaki H, Barrero JM, Li C (2018) Editorial: seed dormancy, germination, and pre-harvest sprouting. Front Plant Sci 9:1783
2. Nakamura S (2018) Grain dormancy genes responsible for preventing pre-harvest sprouting in barley and wheat. Breed Sci 68:295–304
3. Jayakodi M, Padmarasu S, Haberer G et al (2020) The barley pan-genome reveals the hidden legacy of mutation breeding. Nature 588:284–289
4. Walkowiak S, Gao L, Monat C et al (2020) Multiple wheat genomes reveal global variation in modern breeding. Nature 588:277–283
5. Sato K (2020) History and future perspectives of barley genomics. DNA Res 27:dsaa023
6. Takeda K, Hori K (2007) Geographical differentiation and diallel analysis of seed dormancy in barley. Euphytica 153:249–256
7. Nakamura S, Abe F, Kawahigashi H et al (2011) A wheat homolog of mother of FT and TFL1 acts in the regulation of germination. Plant Cell 23:3215–3229

8. Liu S, Sehgal SK, Li J et al (2013) Cloning and characterization of a critical regulator for pre-harvest sprouting in wheat. Genetics 195:263–273
9. Nakamura S, Pourkheirandish M, Morishige H et al (2016) Mitogen-activated protein kinase kinase 3 regulates seed dormancy in barley. Curr Biol 26:775–781
10. Torada A, Koike M, Ogawa T et al (2016) A causal gene for seed dormancy on wheat chromosome 4A encodes a MAP kinase kinase. Curr Biol 26:782–787
11. Sato K, Yamane M, Yamaji N et al (2016) Alanine aminotransferase controls seed dormancy in barley. Nat Commun 7:11625
12. Bayer MM, Rapazote-Flores P, Ganal M et al (2017) Development and evaluation of a barley 50k iSelect SNP array. Front Plant Sci 8:1792
13. Semagn R, Babu R, Hearne S, Olsen M (2014) Single nucleotide polymorphism genotyping using Kompetitive allele specific PCR (KASP): overview of the technology and its application in crop improvement. Mol Breed 33:1–14
14. Iida H (2014) Mugifumi, a beneficial farm work of adding mechanical stress by treading to wheat and barley seedlings. Front Plant Sci 5:453
15. Hisano H, Hoffie RE, Abe F et al (2021) Regulation of germination by targeted mutagenesis of grain dormancy genes in barley. Plant Biotechnol J. https://doi.org/10.1111/pbi.13692
16. Hensel G, Valkov V, Middlefell-Williams J, Kumlehn J (2008) Efficient generation of transgenic barley: the way forward to modulate plant-microbe interactions. J Pant Physiol 165:71–82
17. Hensel G (2020) Genetic transformation of Triticeae cereals-summary of almost three-decade's development. Biotechnol Adv 40:107484
18. Torada A, Ikeguchi S, Koike M (2005) Mapping and validation of PCR-based markers associated with a major QTL for seed dormancy in wheat. Euphytica 143:251–255
19. Torada A, Koike M, Ikeguchi S et al (2008) Mapping of a major locus controlling seed dormancy using backcrossed progenies in wheat (*Triticum aestivum* L.). Genome 51:426–432
20. Bensink L, Koornneef M (2011) Identification and characterization of quantitative trait loci that control seed dormancy in Arabidopsis. Methods Mol Biol 773:165–184
21. Nishimura T, Mori M, Kamada T, Nakane W et al (2019) Detection of QTL for pre-harvest sprouting resistance and grain dormancy in highly sprouting-tolerant wheat. Breed Res 21:115–123
22. Barrero JM, Downie AB, Xu Q, Gubler F (2014) A role for barley CRYPTOCHROME1 in light regulation of grain dormancy and germination. Plant Cell 26:1094–1104
23. Hisano H, Sato K (2016) Genomic regions responsible for amenability to agrobacterium-mediated transformation in barley. Sci Rep 6:37505
24. Hisano H, Meints B, Moscou MJ et al (2017) Selection of transformation-efficient barley genotypes based on TFA (transformation amenability) haplotype and higher resolution mapping of the TFA loci. Plant Cell Rep 36:611–620

Chapter 11

Genome-Wide Association Study of Preharvest Sprouting in Wheat

Jian Yang and Jirui Wang

Abstract

Genome-wide association study (GWAS) is widely used to characterize genes or quantitative trait loci (QTLs) associated with preharvest sprouting and seed dormancy. GWAS can identify both previously discovered and novel QTLs across diverse genetic panels. The high-throughput SNP arrays or next-generation sequencing technologies have facilitated the identification of numerous genetic markers, thereby significantly enhancing the resolution of GWAS. Although various methods have been developed, the fundamental principles underlying these techniques remain constant. Here, we provide a basic technological flow to perform seed dormancy assay, followed by GWAS using population structure control, and compared it with previous identified QTLs and genes.

Key words Seed dormancy, GWAS, QTL, Population structure

1 Introduction

Seed dormancy is an evolutionary adaptation trait that enables plant seeds to maintain sleepy from formation until germination. This property is also a survival mechanism that plants can overcome unfavored environmental conditions to keeping potential ability to initiate a new life cycle. In cereal crops, a reduction in seed dormancy often leads to preharvest sprouting (PHS) damage, which significantly decreases end-use quality and the germination capacity of the seeds in the next sowing [1]. Seed dormancy and PHS resistance are two primary objectives in cereal breeding programs, especially in some grown areas with humid environment.

In the past two decades, the rapid progress in quantitative genetics has led to efficient dissection complex traits, including agronomic traits, quality, abiotic stress, and biotic stress [2–5]. Quantity trait locus (QTL) detected by linkage analysis in segregation population was adopted prior by using two-parent accessions with contrast phenotypes. The adoption of this method

Naoto Kawakami and Kazuhiro Sato (eds.), *Seed Dormancy: Methods and Protocols*, Methods in Molecular Biology, vol. 2830,
https://doi.org/10.1007/978-1-0716-3965-8_11,

is primarily based on extensive screening of phenotype, and several accessions showed talent exhibitions were carried out.

Although this strategy has been successfully applied in genomics studies in wheat, barley, and other cereals, some limitations exist. First, only a small fraction of genetic variants within selective parents can be detected in each population [6–8]. Second, the number of recombinants for each chromosome is typically low, resulting in poor mapping resolution. Additionally, the identification of genetic QTLs often requires a permanent genetic population that needs years to construct.

Based on lower linkage disequilibrium (LD), GWAS has been a powerful tool for functional population genetics analysis, becoming the most exciting strategy that addresses previously unapproachable questions such as mapping variation genome-wide and resolving the genomic basis of a range of phenotypes. In comparison to linkage mapping, GWAS can be shortened cycles of population construction or broken biparents genetic linkage, identifying more genetic variants possible, especially for QTL for complex traits [9, 10]. With this strategy, QTLs previously detected by linkage mapping could be repeatedly identified in widely natural accessions. Moreover, comprehensive studies of QTLs and germplasm can be a reality.

Seed dormancy and germination are controlled by a complex genetic architecture and physiological regulatory mechanisms [11]. Recently, GWAS was applied to study the genetic architecture of seed dormancy (or preharvest sprouting). A panel of 166 common wheat accessions identified 45 loci for seed dormancy [12]. Among these loci, four were associated with *TaMFT-A1* and *R* gene, which were repeatedly detected in linkage analysis. In another panel, 717 Chinese wheat landraces for PHS resistance carried out GWAS using 9740 DArT (Diversity Arrays Technology)-seq and 178,803 SNP markers [13]. GWAS detected three primary QTL: one on 5D that appeared to be novel and two colocated with the grain color transcription factor *TaMyb10* on 3A and 3D (*PHS-3D*), respectively. An additional 32 grain color-related QTLs (GCR-QTLs) were detected when the set of red-grained landraces were analyzed separately. GCR-QTL occurred at high frequencies in the red-grained accessions, and a strong correlation was observed between the number of GCR-QTL and germination rate. These additional factors could be critical for maintaining high levels of PHS resistance and represent targets for introgression into white-grained wheat cultivars. These results also suggested that PHS resistance obtaining in China wheat landraces was due to a combination of natural and artificial selection [13]. The identified QTLs genome was widely showed in gene pyramiding breeding and explaining molecular mechanisms of seed development by gene cloning [13]. GWAS was used to detect the relationship between grain color and PHS in the U.S. winter wheat [14]. Four QTLs for grain color on four chromosomes and 12 QTLs for PHS resistance on ten chromosomes were identified. The QTLs for grain color and PHS resistance were identified on the long arms of the chromosomes 3A and 3D too. Recently, *PHS-3D* was identified from synthetic hexaploid wheat to a 2.4 Mb presence–absence variation region and found that its PHS resistance effect was attributed to the pleiotropic *TaMyb10-D* by enhancing ABA biosynthesis to delay germination in wheat according to integrated omics and functional analyses [15]. At the same time, other QTL on chromosomes 3AS, 3AL, 3B, 4AL, and 7A for PHS resistance showed varied effects under different environments [14]. GWAS based on single- and multi-locus mixed linear models were used to detect PHS loci from 192 wheat varieties [16]. The single-locus model identified 23 loci for PHS ($P < 0.0001$) and explained 6.0%–18.9% of the phenotypic variance. Twenty loci were consistent with known QTLs. Three single-nucleotide polymorphism markers closely linked with three major loci on chromosomes 1AL, 3BS, and 6BL, respectively. Moreover, PHS and Falling Numbers (FN) could be detected with the same panel by GWAS [17]. However, there was no overlap between the QTL detected based on FN and PHS, and there was little correlation between the two traits.

2 Materials

2.1 Evaluation of Grain Dormancy

1. Enough number of harvest-ripe grains.
2. Petri dishes.
3. Filter paper.
4. Temperature-controlled greenhouse or incubator.
5. Sterilized water.
6. Refrigerator with temperature to −20 °C.
7. Gibberellin.
8. Any fresh tissue of the plant.

2.2 Hardware and Software

1. Computer (preferably Intel Core i7 or better processor and at least 8 GB memory).
2. STRUCTURE harvest online program [18] (http://taylor0.biology.ucla.edu/structureHarvester/; https://github.com/dentearl/structureHarvester/).
3. TASSEL software [19] (*see* **Note 1**).
4. R software (www.r-project.org/).
5. CMplot (https://cran.r-project.org/web/packages/CMplot/).

2.3 Database

1. Plant genome sequence database (http://plants.ensembl.org/info/data/ftp/index.html).

3 Methods

3.1 Evaluation of Preharvest Sprouting Resistance

1. Harvest wheat spikes upon reaching physiological maturity and subsequently air-dry them at room temperature for a period of 7 days (*see* **Note 2**). During this drying phase, carefully hand-thresh a sufficient quantity of grains, taking special care to avoid any damage to the seed coat or the embryos.
2. After 7 days, store threshed grains at −20 °C (usually in a freezer) until needed for the experiment.
3. Remove the grains from the freezer and count out 50 grains for each replication per accession. Place these grains on a filter paper in a Petri dish (*see* **Note 3**).
4. Add about 10 mL of sterilized water to the Petri dish and keep it at 24 °C under normal light conditions (Fig. 1).
5. Pick out germinated grains and record the number daily. Germination is defined as the rupture of the grain coat by the emerging radicle (*see* **Note 4**).
6. Incubate for 7 days and keep replenishing water in Petri dish to avoid drying.
7. Treat with 0.05% gibberellin for 3 days in the dark. After that, the number of ungerminated grains was recorded and removed for the total number of grains for germination assay.

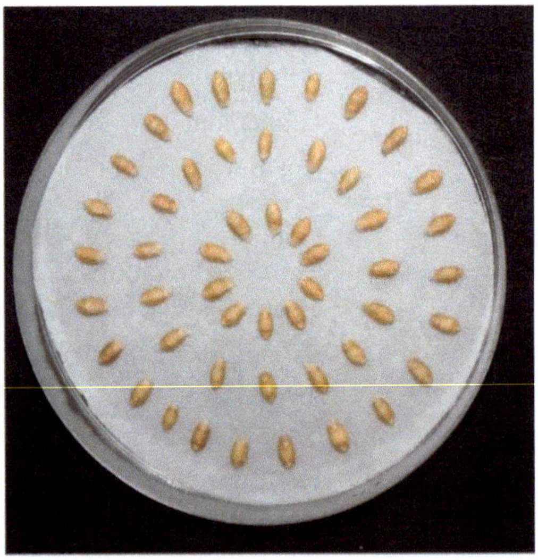

Fig. 1 Germination assay of wheat accession

8. Calculate the GI value and GR value to define the dormancy character. The GI was calculated by $GI = \frac{7 \times n_1 + 6 \times n_2 + 5 \times n_3 + 4 \times n_4 + 3 \times n_5 + 2 \times n_6 + 1 \times n_1}{7 \times n}$, where n_1, n_2,... n_7 mean germinated grains in first, second,... 7th day, and n is the total number of grains used for germination assay [20].

3.2 DNA Extraction and Genotyping

1. Fresh tissue (e.g., leaf, seedling) can be used for DNA extraction using the CTAB method [21].
2. The qualified DNA can be genotyped by high-throughput genotype in multiple platforms (Table 1).

3.3 Population Structure

1. Genotyping data with appropriate format (one line with one accession, two alleles encoded "0" or "1" for each accession in the two columns, respectively). The accession ID, which is labeled in the first column, and marker ID, which is in the first row, were uploaded into TESSAL software [19]. An admixture model with 10,000 burn-in iterations, followed by 10,000 Markov chain Monte Carlo iterations, for accurate parameter estimation, was used with five independent replicates for each K value. Generally, the K was ranging from $K = 1$ to $K = 10$.
2. Collect all result files in the working directory to a package (.zip) and upload them into Structure Harvester online program and click the "Harvest!" button. The K with the highest value of Delta K is generally determined as the subpopulation classification.
3. The "Inferred ancestry of individuals" is Q matrix file and referred to population structure for the following use.

3.4 Genome-Wide Association Study

1. Association can be performed using Mixed Linear Model (MLM) implemented in TASSEL software. Files including genotype (format can be seen in the instruction of TASSEL), phenotype, and Q matrix were accessed into the software.

Table 1
High-throughput genotyping platforms

Technology	Method	Most used productions
SNP array	DNA hybridization	Illumina wheat 9 K iSelect SNP array [22]; Illumina wheat 90 K iSelect SNP genotyping array [23]; axiom® wheat 660 K SNP array; wheat 55 K SNP array; axiom® HD wheat genotyping (820 K) array [24]; 1536-SNP barley SNP array [25].
GBS	Restriction enzyme digestion	1351 diversity array technology (DArT) markers [26], GBS [27], RAD [28].
NGS	Sequence	Resequence [29], Exome capture [30–32]

Using filter function, the allele frequency of SNPs less than 5% can be removed from the analysis.

2. Calculate pairwise kinship through "Analysis-Relatedness-Kinship."
3. Initiate genome-wide association mapping by integrating genotype data, phenotype observations, kinship information, and population structure analysis. Consider SNPs with a *P* value less than 0.001 as significantly associated with seed dormancy.
4. Output data with SNP ID, chromosome name, physical position, and *P* value. Use CMplot under R environment to Manhattan plot.

```
--------------------------
Install packages ("CMplot")
library(CMplot)
data<-("GWAS.txt",header=T)
CMplot(data,plot.type="m",threshold=0.001,threshold.col=c
("red"),col=c("#377EB8","#4DAF4A","#984EA3"),threshold.lty=c
(1,2),threshold.lwd=c(1,1),amplify=F,signal.cex=c(1,1),sig-
nal.pch = c(20,20),signal.col = c("red","orange"),file="pdf")
--------------------------
```

3.5 Candidate Gene Prediction

Several genes implicated in seed dormancy and preharvest sprouting resistance have been successfully cloned, as detailed in Table 2. To locate these genes in the reference genome, their sequences (or IDs) can be retrieved from relevant publications and then subjected to BLAST analysis on the designated website.

1. Once the genome sequence is downloaded, a local database will be constructed by NCBI blast+ package (National Center for Biotechnology Information, Bethesda, MD, USA). Download the blast+ package at ftp://ftp.ncbi.nlm.nih.gov/blast/executables/blast+/LATEST/. Unzip the package and install it on a suitable disk (e.g., disk C) of your computer.

```
--------------------------
cd C:\blast-2.2.25+\db
blastdbcmd -in ChineseSpringV1.fasta -dbtype nucl -out Chi-
neseSpringV1db
blastn -query Vp1.fasta -db ChineseSpringV1db -out Vp1_
ChineseSpringV1.txt -outfmt 7
--------------------------
```

Table 2
Some key genes controlling seed dormancy

Gene name	Species	Traits	Gene ontology	Reference
Vp1	Wheat	Preharvest sprouting	B3 domain transcriptional factor	[33]
MFT-3A	Wheat	Seed dormancy	Homolog of FLOWERING LOCUS T (FT) and TERMINAL FLOWER 1 (TFL1)	[34, 35]
PM19-A1; PM19-A2	Wheat	Seed dormancy	ABA-induced wheat plasma membrane 19 family	[36]
MKK3	Wheat	Seed dormancy	Mitogen-activated protein kinase kinase 3	[37]
MKK3	Barley	Seed dormancy	Mitogen-activated protein kinase kinase 3	[38]
AlaAT	Barley	Seed dormancy	Alanine aminotransferase	[39]
PHS-3D/ TaMyb10-D	Wheat	Preharvest sprouting, grain color	MYB domain transcriptional factor	[15], [40]
Sdr4	Rice	Seed dormancy	Proteins with unknown functions	[41]
DOG1	Arabidopsis	Seed dormancy	Plant specific protein family with a domain shared by the D class bZIP DNA-binding proteins	[42]

2. The physical position of SNP probes can also be acquired by using their sequences as the query. Comparison of the positions of associated SNPs and seed dormancy-related genes will provide primary evidence to indicate whether QTLs were identified or not.
3. Furthermore, associated markers can be converted to KASP markers for breeding [43].

4 Notes

1. Many packages can be used to conduct GWAS, including GAPIT, EMMAX, PLINK. Here, TASSEL is a graphical interface-based software.
2. Ensure proper ventilation, and avoid storing grains direct sunlight or at high temperatures.
3. If the work cannot be completed in one day, the grains of uncompleted lines should be stored at −20 °C until further processing. The grains stored at −20 °C can retain dormancy for an extended period.
4. Sometimes, the grains may become mildewed. Pick them out and remove them from the total count of grains.

References

1. Buchanan AM, Nicholas EM (1980) Sprouting, alpha-amylase and breadmaking quality. Cereal Res Commun 8:23–28
2. Asif MA, Schilling RK, Tilbrook J et al (2018) Mapping of novel salt tolerance QTL in an Excalibur × kukri doubled haploid wheat population. Theor Appl Genet 131:2179–2196
3. Wang RX, Hai L, Zhang XY et al (2009) QTL mapping for grain filling rate and yield-related traits in RILs of the Chinese winter wheat population Heshangmai×Yu8679. Theor Appl Genet 118:313–325
4. Schnurbusch T, Paillard S, Schori A et al (2004) Dissection of quantitative and durable leaf rust resistance in Swiss winter wheat reveals a major resistance QTL in the Lr34 chromosomal region. Theor Appl Genet 108:477–484
5. Zhang W, Chao F, Manthey F et al (2008) QTL analysis of pasta quality using a composite microsatellite and SNP map of durum wheat. Theor Appl Genet 117:1361–1377
6. Cabral AL, Jordan MC, McCartney CA et al (2014) Identification of candidate genes, regions and markers for pre-harvest sprouting resistance in wheat (*Triticum aestivum* L.). BMC Plant Biol 14:340
7. Chen C-X, Cai S-B, Bai G-H (2008) A major QTL controlling seed dormancy and pre-harvest sprouting resistance on chromosome 4A in a Chinese wheat landrace. Mol Breed 21:351–358
8. Yang J, Tan C, Lang J et al (2019) Identification of *qPHS.Sicau-1B* and *qPHS.Sicau-3D* from synthetic wheat for pre-harvest sprouting resistance wheat improvement. Mol Breed 39: 132
9. Huang X, Han B (2014) Natural variations and genome-wide association studies in crop plants. Annu Rev Plant Biol 65:531–551
10. Korte A, Farlow A (2013) The advantages and limitations of trait analysis with GWAS: a review. Plant Methods 9:29
11. Graeber K, Nakabayashi K, Miatton E et al (2012) Molecular mechanisms of seed dormancy. Plant Cell Environ 35:1769–1786
12. Zuo J, Lin C-T, Cao H et al (2019) Genome-wide association study and quantitative trait loci mapping of seed dormancy in common wheat (*Triticum aestivum* L.). Planta 250: 187–198
13. Zhou Y, Tang H, Cheng M-P et al (2017) Genome-wide association study for pre-harvest sprouting resistance in a large germplasm collection of Chinese wheat landraces. Front Plant Sci 8:401
14. Lin M, Zhang D, Liu S et al (2016) Genome-wide association analysis on pre-harvest sprouting resistance and grain color in U.S. winter wheat. BMC Genomics 17:794
15. Lang J, Fu Y, Zhou Y et al (2021) *Myb10-D* confers *PHS-3D* resistance to pre-harvest sprouting by regulating *NCED* in ABA biosynthesis pathway of wheat. New Phytol 230: 1940–1952
16. Zhu Y, Wang S, Wei W et al (2019) Genome-wide association study of pre-harvest sprouting tolerance using a 90K SNP array in common wheat (*Triticum aestivum* L.). Theor Appl Genet 132:2947–2963
17. Martinez SA, Godoy J, Huang M et al (2018) Genome-wide association mapping for tolerance to Preharvest sprouting and low falling numbers in wheat. Front Plant Sci 9:141
18. Earl DA, vonHoldt BM (2012) STRUCTURE HARVESTER: a website and program for visualizing STRUCTURE output and implementing the Evanno method. Conserv Genet Resour 4:359–361
19. Bradbury PJ, Zhang Z, Kroon DE et al (2007) TASSEL: software for association mapping of complex traits in diverse samples. Bioinformatics 23:2633–2635
20. Walker-Simmons M (1987) ABA levels and sensitivity in developing wheat embryos of sprouting resistant and susceptible cultivars. Plant Physiol 84:61–66
21. Gupta P, Salava H, Sreelakshmi Y et al (2020) A low-cost high-throughput method for plant genomic DNA isolation. Methods Mol Biol 2072:1–7
22. Cavanagh CR, Chao S, Wang S et al (2013) Genome-wide comparative diversity uncovers multiple targets of selection for improvement in hexaploid wheat landraces and cultivars. Proc Natl Acad Sci 110:8057–8062
23. Wang S, Wong D, Forrest K et al (2014) Characterization of polyploid wheat genomic diversity using a high-density 90,000 single nucleotide polymorphism array. Plant Biotechnol J 12:787–796
24. Winfield MO, Allen AM, Burridge AJ et al (2016) High-density SNP genotyping array for hexaploid wheat and its secondary and tertiary gene pool. Plant Biotechnol J 14:1195–1206

25. Close TJ, Bhat PR, Lonardi S et al (2009) Development and implementation of high-throughput SNP genotyping in barley. BMC Genomics 10:582
26. Akbari M, Wenzl P, Caig V et al (2006) Diversity arrays technology (DArT) for high-throughput profiling of the hexaploid wheat genome. Theor Appl Genet 113:1409–1420
27. Poland JA, Brown PF, Sorrells ME, Jnnink J-L (2012) Development of high-density genetic maps for barley and wheat using a novel two-enzyme genotyping-by-sequencing approach. PLoS One 7:e32253
28. Chutimanitsakun Y, Nipper RW, Cuesta-Marcos A et al (2011) Construction and application for QTL analysis of a restriction site associated DNA (RAD) linkage map in barley. BMC Genomics 12:4
29. Hao C, Jiao C, Hou J et al (2020) Resequencing of 145 landmark cultivars reveals asymmetric sub-genome selection and strong founder genotype effects on wheat breeding in China. Mol Plant 13:1733–1751
30. Krasileva K et al. (2014) Exome capture and TILLING in tetraploid and Hexaploid wheats. In International Plant and Animal Genome Conference XXII 2014
31. Dong C, Zhang L, Chen Z et al (2020) Combining a new exome capture panel with an effective varBScore algorithm accelerates BSA-based gene cloning in wheat. Front Plant Sci 11:1249
32. Liu F, Zhao Y, Beier S et al (2020) Exome association analysis sheds light onto leaf rust (*Puccinia triticina*) resistance genes currently used in wheat breeding (*Triticum aestivum* L.). Plant Biotechnol J 18:1396–1408
33. Mckibbin RS, Wilkinson MD, Bailey PC et al (2002) Transcripts of VP-1 homeologues are misspliced in modern wheat and ancestral species. Proc Natl Acad Sci 99:10203–10208
34. Nakamura S, Abe F, Kawahigashi H et al (2011) A wheat homolog of MOTHER OF FT AND TFL1 acts in the regulation of germination. Plant Cell 23:3215–3229
35. Liu S, Sehgal SK, Li J et al (2013) Cloning and characterization of a critical regulator for pre-harvest sprouting in wheat. Genetics 195:263–273
36. Barrero JM, Cananagh C, Verbyla KL et al (2015) Transcriptomic analysis of wheat near-isogenic lines identifies *PM19-A1* and *A2* as candidates for a major dormancy QTL. Genome Biol 16:93
37. Torada A, Koike M, Ogawa T et al (2016) A causal gene for seed dormancy on wheat chromosome 4A encodes a MAP kinase kinase. Curr Biol 26:782–787
38. Nakamura S, Pourkheirandish M, Morishige H et al (2016) Mitogen-activated protein kinase kinase 3 regulates seed dormancy in barley. Curr Biol 26:775–781
39. Sato K, Yamane M, Yamaji N et al (2016) Alanine aminotransferase controls seed dormancy in barley. Nat Commun 7:11625
40. Himi E, Maekawa M, Miura H, Noda K (2011) Development of PCR markers for *Tamyb10* related to *R-1*, red grain color gene in wheat. Theor Appl Genet 122:1561–1576
41. Sugimoto K, Takauchi Y, Ebana K et al (2010) Molecular cloning of *Sdr4*, a regulator involved in seed dormancy and domestication of rice. Proc Natl Acad Sci USA 107:5792–2597
42. Bentsink L, Jowett J, Hanhart CJ, Koornneef M (2006) Cloning of *DOG1*, a quantitative trait locus controlling seed dormancy in *Arabidopsis*. Proc Natl Acad Sci USA 103:17042–17047
43. Rasheed A, Wen W, Gao F et al (2016) Development and validation of KASP assays for genes underpinning key economic traits in bread wheat. Theor Appl Genet 129:1843–1860

Chapter 12

Functional Analysis of Seed Dormancy Genes by Biolistic Transient Gene Expression in Immature Embryos of Wheat

Fumitaka Abe, Masahiko Mori, and Shingo Nakamura

Abstract

Seed dormancy genes typically suppress germination and cell division. Therefore, overexpressing these genes can negatively affect tissue culture, interfering with the generation of transgenic plants and thus hampering the analysis of gene function. Transient expression in target cells is a useful approach for studying the function of seed dormancy genes. Here, we describe a protocol for transiently expressing genes related to seed dormancy in the scutellum of immature wheat (*Triticum aestivum*) embryos to analyze their effects on germination.

Key words Transient gene expression, Immature embryo, Biolistic delivery, Precocious germination, Wheat, Tissue culture, *Mother of FT and TFL1*

1 Introduction

Genetic transformation techniques are indispensable for gene functional analyses, especially in wheat (*Triticum aestivum* L.) with its complex allohexaploid genome [1]. For example, genetic transformation is widely used for complementation tests with genomic fragments, and for the production of transformants with altered gene expression levels or patterns, using overexpression, promoter swapping, or RNA interference methods. With the exception of *Agrobacterium* infection of floral organs in *Arabidopsis*, referred to as the floral dip method, most methods produce transformed plants through tissue culture. This has provided key insight into the functions of many genes, but poses a problem for genes that affect key processes such as cell division, because tissue culture-based methods will not produce transformants overexpressing a gene that interferes with tissue culture. As an alternative to stable genetic transformation, transient gene expression in target cells is a useful approach for studying the functions of genes that interfere with generating transgenic plants.

Naoto Kawakami and Kazuhiro Sato (eds.), *Seed Dormancy: Methods and Protocols*, Methods in Molecular Biology, vol. 2830,
https://doi.org/10.1007/978-1-0716-3965-8_12,

Many of the genes that regulate seed development can also interfere with the generation of transgenic plants. Seed development involves embryo development followed by seed maturation. At the end of embryo development, cell division in the embryo is arrested, and embryo dormancy is initiated to prevent embryo growth during seed maturation [2]. After seed maturation, the seed dormancy program prevents seed germination and the accompanying cell division. Understanding the functions of dormancy-related genes has key implications for assuring timely seed germination in the field, and for preventing precocious germination of grain.

Since dormancy-related genes suppress cell division, obtaining transformants overexpressing these genes is difficult. We observed such difficulties during the production of transformants overexpressing the *Delay of Germination* (*DOG1*) gene [3]. *TaDOG1L4*, which has a large effect in promoting seed dormancy, suppressed callus growth during tissue culture, resulting in a transformation efficiency clearly lower than that of the barley (*Hordum vulgare*) gene *HvDOG1L1*, which has a relatively small effect on seed dormancy. Similarly, during transformation of wheat with an overexpression construct of the wheat homolog of *MOTHER OF FT AND TFL* (*TaMFT*) gene, which was identified by microarray analysis as being highly expressed in dormant embryos, callus did not grow at all and transformed wheat could not be obtained (unpublished data). In response to this result, we developed a method to transiently express the *TaMFT* gene in immature embryos to analyze its effect on germination [4].

Here, we describe our procedure for transiently expressing a transgene in the scutellum of immature wheat embryos to analyze its effect on precocious germination. In general, immature embryos germinate precociously when extracted from immature seeds and placed on a suitable medium. Thus, we exploit this phenomenon to evaluate the *MFT* effect on germination by examining whether the isolated immature embryos can germinate or not when *TaMFT* driven by ubiquitin promoter overexpression construct is delivered to the scutellum of immature embryos using particle bombardment and subsequently placed the embryos onto germination medium. Our experiments showed that the immature embryos delivered by the construct developed to a certain stage but did not germinate precociously even after incubation of 10 days. In contrast, as control, when we delivered one base substituted construct to create a stop codon to the immature embryos in the same way, we observed almost of all immature embryos germinated precociously after incubation of 7 days. Furthermore, embryos whose precocious germination was suppressed by transient expression of the *TaMFT* gene germinated immediately when transferred onto medium containing gibberellin, but they did not germinate in the absence of gibberellin. This procedure clearly shows that the function of *TaMFT* is a suppressor for germination.

2 Materials

2.1 Preparation of Immature Embryos

1. Wheat plants for obtaining immature embryos: Sow six seeds of wheat cultivar "Bobwhite" line SH9826 [5] in 18-cm plastic pots containing 2.6 L of a 2:1 mixture of Sakata Supermix A (fine peat moss) and Nippi prefertilized and granulated soil (Nihon Hiryo) with 2.6 g of the controlled release fertilizer Osmocote Exact Mini 3-4 M (15-9-11 + 2MgO + TE). Grow plants in a glasshouse with day/night temperatures of 16 °C/ 10 °C under an 8-h light/16-h dark photoperiod. Just before heading (10–12 weeks depending on season), transfer plants to a controlled environmental chamber with adjusted day/night temperatures of 20 °C/13 °C under a 14-h light (300–500 μmol/m^2/s)/10-h dark photoperiod.
2. 70% (v/v) ethanol, 0.5% (v/v) sodium hypochlorite, and sterilized distilled water.
3. A stereoscopic microscope in laminar flow clean bench.
4. Osmotic medium: Add 2.3 g MS basic salts, 0.5 mL 1000× modified MS vitamin stock, and 200 mg MES to 600 mL distilled water and adjust pH to 5.8. Make the volume up to 700 mL, add 2 g Phytagel, and autoclave at 121 °C for 20 min. Cool to 50 °C before adding 300 mL of 0.5 g/mL maltose. Pour 20 mL of medium into each 90-mm Petri dish. Store at room temperature (RT) in the dark.

2.2 Particle Bombardment and Germination Test

1. Gold particles: Weigh 40 mg gold particles (1.0 μm in diameter) in a 2.0-mL centrifuge tube and sterilize by washing three times with ethanol. Discard the supernatant by brief centrifugation, and resuspend pellet in 1 mL of sterilized distilled water.
2. 2.5 M $CaCl_2$ and 100 mM spermidine.
3. Plasmid DNA (1 μg/μL) harboring overexpression cassette for the gene of interest (*see* **Note 1**).
4. Modified Murashige and Skoog (MS) vitamin stock (1000×): 100 g/L *myo*-inositol, 2 g/L glycine, 1 g/L thiamine hydrochloride, 0.5 g/L pyridoxine hydrochloride, and 0.5 g/L nicotinic acid. Store in 1-mL aliquots at −20 °C.
5. Maltose (0.5 g/mL): Dissolve 500 g maltose in 600 mL distilled water. Make up the volume to 1000 mL and filter sterilize using a 0.22-μm filter. Store at 4 °C.
6. Osmotic medium described in Subheading 2.1, **step 4**.

7. Germination medium: Add 2.3 g MS basic salts, 0.5 mL 1000× modified MS vitamin stock, 200 mg MES, and 30 g maltose to 900 mL distilled water and adjust pH to 5.8. Make up the volume to 1000 mL, add 8 g agar, and autoclave at 121 °C for 20 min. Pour 20 mL of medium into each 90-mm petri dish. Store at room temperature in the dark.
8. Stereoscopic microscope in laminar flow clean bench.
9. Ultrasonic cleaner.
10. PDS1000/He particle bombardment system (BioRad).

3 Methods

3.1 Isolation of Immature Embryos

1. Collect immature seeds of the correct developmental stage from panicles at approximately 11 days after anthesis (*see* **Note 2**). Sterilize immature seeds with 70% ethanol for 1 min followed by 0.5% sodium hypochlorite for 15 min and then wash three times with sterilized distilled water.
2. Isolate immature embryos from immature seeds under a stereoscopic microscope in a laminar flow clean bench. Transfer immature embryos to osmotic medium. Place about 20 embryos in a circle of 1.2 cm diameter in the center of each plate, scutellum side up (Fig. 1) (*see* **Note 3**).
3. Incubate plates at 25 °C for at least 3–4 h before bombardment.

3.2 Particle Bombardment

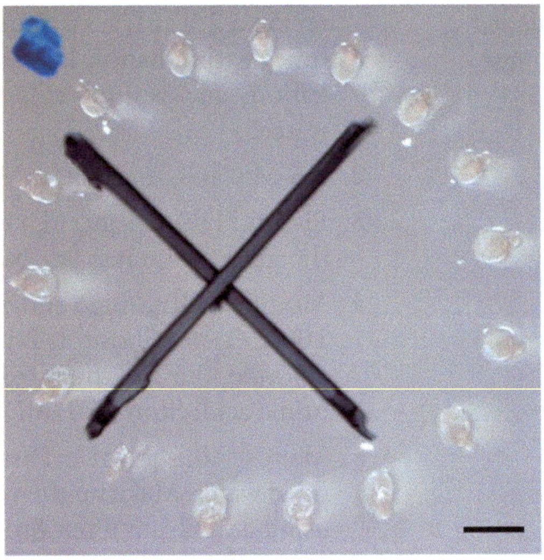

Fig. 1 Immature embryos before bombardment. Scale bar, 2 mm

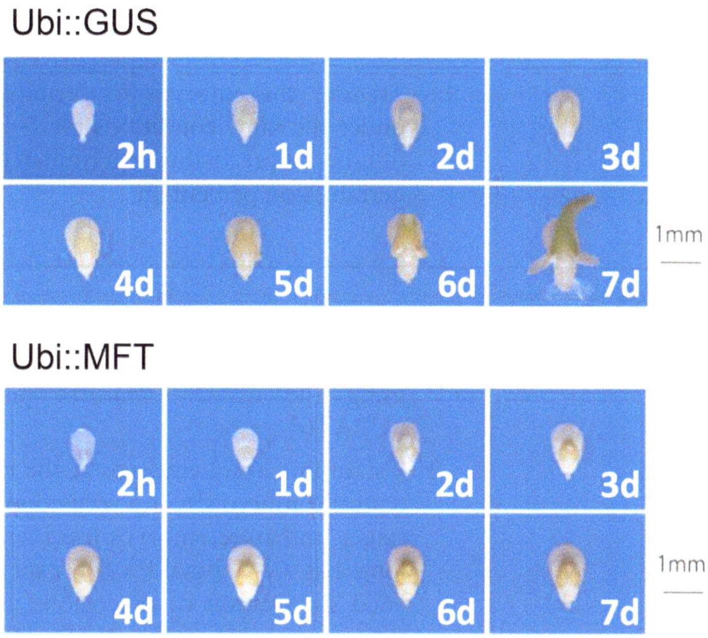

Fig. 2 Time course of isolated immature embryo development after transformation with Ubi::MFT or Ubi::GUS.h; hours after transformation, d; days after transformation. Scale bar, 1 mm

1. Add 2 mg gold particles (50 μL) to a 2.0-mL microcentrifuge tube and sonicate (40 kHz) using an ultrasonic cleaner for 1 min.
2. Add 5 μg of plasmid DNA (5 μL) and mix thoroughly by pipetting.
3. Mix 50 μL of 2.5 M $CaCl_2$ and 20 μL of 100 mM spermidine in the tube lid. Close the lid and immediately vortex for 5 min. Allow mixture to settle at RT for 3 min.
4. Centrifuge at 12,000 rpm for 10 s to collect the gold particles. Remove the supernatant without disturbing the pellet.
5. Add 150 μL ethanol and resuspend. Centrifuge at 12,000 rpm for 10 s to wash the particles.
6. Repeat the wash twice more and resuspend in 100 μL ethanol.
7. Place a macro-carrier into the macro-carrier holder.
8. Spread 5 μL of gold particle suspension onto the central region of the macro-carrier. Air dry in a laminar flow clean bench.
9. After drying, bombard the immature embryos using a PDS1000/He particle delivery system according to the manufacturer's instructions. Set up the equipment with 900 psi helium pressure and a target distance of 5.5 cm from the stopping screen.

10. Incubate bombarded immature embryos at 25 °C in the dark for 1 day.
11. Transfer the embryos to germination medium and culture under the same conditions for 7–10 days (Fig. 2). Count the number of precociously germinated embryos daily to calculate germination percentage.

4 Notes

1. We have succeeded in transiently expressing the *MFT* gene using the maize *Ubiquitin* promoter from the vector pUBA [6].
2. Use of immature embryos at the right developmental stage is critical. Embryo size is a very good indicator of developmental stage. Immature embryos less than 1 mm in length along the long axis are optimal for this experiment. We observed weakened suppression of precocious germination following transient expression of the *MFT* gene when immature embryos larger than 1 mm were used.
3. Immature embryos should be handled with care so as not to damage them. Since gold particles tend to concentrate in the center of the plate, we usually place the embryo axis outward to avoid damage from excess gold particles.

References

1. Hensel G (2020) Genetic transformation of Triticeae cereals – Summary of almost three-decade's development. Biotechnol Adv 40: 107484
2. Raz V, Bergervoet JHW, Koornneef M (2001) Sequential steps for developmental arrest in *Arabidopsis* seeds. Development 128:243–252
3. Ashikawa I, Mori M, Nakamura S, Abe F (2014) A transgenic approach to controlling wheat seed dormancy level by using Triticeae *DOG1*-like genes. Transgenic Res 23:621–629
4. Nakamura S, Abe F, Kawahigashi H, Nakazono K, Tagiri A, Matsumoto T, Utsugi S, Ogawa T, Handa H, Ishida H, Mori M, Kawaura K, Ogihara Y, Miura H (2011) A Wheat Homolog of MOTHER OF FT AND TFL1 Acts in the Regulation of Germination. Plant Cell 23:3215–3229
5. Pellegrineschi A, Noguera LM, Skovmand B, Brito RM, Velazquez L, Salgado MM, Hrnandez R, Warburton M, Hoisington D (2002) Identification of highly transformable wheat genotypes for mass production of fertile transgenic plants. Genome 45:421–430
6. Toki S, Takamatsu S, Nojiri C, Ooba S, Anzai H, Iwata M, Christensen AH, Quail PH, Uchimiya H (1992) Expression of a maize ubiquitin gene promoter-*bar* chimeric gene in transgenic rice. Plant Physiol 100:1503–1507

Chapter 13

Genome Editing to Produce Knockout Mutations of Seed Dormancy Genes in Wheat

Fumitaka Abe, Yoko Kamiya, Yuji Ishida, Hiroshi Hisano, Kanako Kawaura, Toshihiko Komari, and Kazuhiro Sato

Abstract

Knockout mutants provide definitive information about the functions of genes related to agronomic traits, including seed dormancy. However, it takes many years to produce knockout mutants using conventional techniques in polyploid plants such as hexaploid wheat. Genome editing with sequence-specific nucleases is a promising approach for obtaining knockout mutations in all targeted homoeologs of wheat simultaneously. Here, we describe a procedure to produce a triple recessive mutant in wheat via genome editing. This protocol covers the evaluation of gRNA and *Agrobacterium*-mediated transformation to obtain edited wheat seedlings.

Key words Genome editing, CRISPR/Cas9, Immature embryo, *Agrobacterium*-mediated transformation, Wheat

1 Introduction

The emergence of programmable sequence-specific nucleases (SSNs) represents a major breakthrough in the functional analysis of genes. These precise, efficient tools are used for genome editing to produce specific mutations, including knockout mutations [1, 2]. SSN techniques include those using zinc-finger nucleases, transcription activator-like effector nucleases (TALENs), and clustered regularly interspaced short palindromic repeats (CRISPR)/CRISPR-associated protein 9 (Cas9). Since SSNs can introduce mutations into multiple target genes simultaneously based on sequence identity, genome editing using SSNs is highly effective for producing knockout mutants, particularly in polyploid species such as hexaploid wheat (*Triticum aestivum* L.). In an early study, Wang et al. [3] used TALEN to introduce mutations into all three homoeologs of the wheat *MLO* (*MILDEW-RESISTANCE LOCUS O*) gene to confer powdery mildew resistance.

Naoto Kawakami and Kazuhiro Sato (eds.), *Seed Dormancy: Methods and Protocols*, Methods in Molecular Biology, vol. 2830, https://doi.org/10.1007/978-1-0716-3965-8_13,

The CRISPR/Cas9 system consists of the Cas9 endonuclease in complex with a guide RNA (gRNA). The gRNA is engineered so that a 20-nucleotide sequence at its 5′ end is homologous to the target site(s) for genome editing. CRISPR/Cas9 is currently the most widely used genome-editing system because of the ease of vector construction and its applicability to a wide range of species. This system has increasingly been employed in wheat [4]. Biolistic transformation was initially used to introduce CRISPR/Cas9 into wheat cells due to its high efficiency. A few years later, *Agrobacterium*-mediated transformation of the CRISPR/Cas9 system into wheat was successfully performed [4].

Mutants produced by genome editing are useful for studying seed dormancy. Using the CRISPR/Cas9 system, we recently developed a loss-of-function triple mutant of *TaQsd1*, an ortholog of a gene that controls seed dormancy in barley, resulting in longer seed dormancy in wheat [5]. By combining the use of genetic information from barley (a diploid crop closely related to wheat) with a genome-editing technique that can simultaneously modify all homoeologs, we produced wheat plants with reduced preharvest sprouting much more quickly than would have been possible using conventional techniques.

The choice of a suitable gRNA sequence is a critical factor for efficient CRISPR/Cas9-mediated genome editing. Therefore, given that wheat transformation requires a large investment in time and resources, it is important to evaluate the efficiency of the gRNA sequence before using it for transformation. This is often done using the protoplast assay system [6, 7]. However, since it is difficult to regenerate wheat plants from protoplasts, that requires preparing a separate wheat transformation system in addition to the protoplast culture used for testing. As an alternative, we recently developed a gRNA evaluation method using immature wheat embryos, which serve as explants for *Agrobacterium*-mediated transformation [8].

Here, we describe our protocol for producing triple knockout mutants by CRISPR/Cas9-mediated gene editing in hexaploid wheat. The present protocol consists of three major steps. The first step is to design the gRNA of the target gene and construct a vector for CRISPR/Cas9. The second step is to evaluate the designed gRNA using a biolistic transient assay system. The final step is to transform the constructed vector into wheat by the *Agrobacterium*-mediated method.

2 Materials

2.1 Vector Construction

1. Plasmids for CRISPR/Cas9: pU6gRNA and pZH_OsU6gRNA_PubiMMCas9 [9].

2. Forward and reverse DNA oligonucleotides of gRNA. These oligos contain the identified 20-nucleotide sequence and must include overhangs specific for the BbsI-digested gRNA cloning vector. For example, for pU6gRNA-oligo and TaQsd1 target_1 (5′-ACGGATCCACCTCCCTGCAG-3′, Abe et al. 2019), the following oligos were used:

   5′-GTTGACGGATCCACCTCCCTGCAG-3′

   3′-TGCCTAGGTGGAGGGACGTCCAAA-5′
3. Restriction enzymes: *Bbs*I, *Asc*I, and *Pac*I.
4. Gel electrophoresis apparatus.
5. Agarose.
6. Gel extraction kit.
7. Ligation kit.
8. Competent *Escherichia coli* cells.
9. Plasmid DNA extraction kit.
10. OsU6-2F primer (5′--TGCTGGAATTGCCCTTGGATCATGAACCAA-3′) for sequencing the plasmid.
11. Sanger sequencing kit.
12. LB liquid broth and agar plates (plus 50 mg/L ampicillin or 75 mg/L spectinomycin).
13. *Agrobacterium tumefaciens* strain EHA101 [10].

2.2 Biolistic Transient Assays to Evaluate gRNA

1. 1000× modified MS vitamin stock: 100 g/L myoinositol, 2 g/L glycine, 1 g/L thiamine hydrochloride, 0.5 g/L pyridoxine hydrochloride, and 0.5 g/L nicotinic acid in distilled water. Store in 1 mL aliquots at −20 °C.
2. Embryo collection medium (WMS-liq): Add 0.46 g MS basic salts, 0.1 mL of 1000× modified MS vitamin stock, 10 g glucose, and 0.5 g 2-(*N*-morpholino)ethanesulfonic acid (MES) to 900 mL distilled water. Adjust the pH to 5.8. Add more distilled water to a final volume of 1000 mL, filter sterilize using a 0.22-μm filter, and store at 4 °C.
3. 0.5 g/mL Maltose solution: Add 500 g maltose to 600 mL distilled water. Add distilled water to 1000 mL, filter sterilize using a 0.22-μm filter, and store at 4 °C.
4. Osmotic medium (WMS-Osm): Add 4.6 g MS basic salts, 1 mL of 1000× modified MS vitamin stock, and 150 mg L-asparagine to 600 mL distilled water and adjust the pH to 5.8. Add distilled water to 700 mL, add 8 g agar, and autoclave at 121 °C for 20 min. Cool to 50 °C and add 300 mL of 0.5 g/mL maltose and 2.5 mL of 1 mg/mL 2,4-dichlorophenoxyacetic acid (2,4-D). Pour 20 mL of

medium into each 90-mm Petri dish. Store at room temperature (RT) in the dark.

5. Plasmids: pU6gRNA containing a gRNA expression cassette and pE(R4-R3)ZmUbi_OsCas9_ver.3 containing a Cas9 expression cassette [11].
6. Stereoscopic microscope.
7. Gold particles: Weigh 20 mg gold particles (0.6 μm in diameter) into a 2.0-mL centrifuge tube and sterilize by washing three times with ethanol. Discard the supernatant by centrifuging and resuspend in 1 mL sterile ddH_2O.
8. PDS1000/He particle bombardment system (Bio-Rad).
9. Callus induction medium (WMS-CI): Add 4.6 g MS basic salts, 1 mL of 1000× modified MS vitamin stock, 30 g sucrose, and 150 mg L-asparagine to 900 mL distilled water, and adjust the pH to 5.8. Add distilled water to 1000 mL, add 8 g agar, and autoclave at 121 °C for 20 min. Cool to 50 °C and add 2.5 mL of 1 mg/mL 2,4-D. Pour 20 mL of medium into each 90-mm Petri dish. Store at room temperature (RT) in the dark.
10. DNA extraction buffer: Add 200 mL of 1 M Tris–HCl, pH 8.0, 14.61 g NaCl, and 50 mL of 0.5 M EDTA to 700 mL distilled water. Add distilled water to 1000 mL and autoclave at 121 °C for 20 min.
11. TissueLyser II (Qiagen).
12. TA Cloning Kit.

2.3 Agrobacterium-Mediated Transformation

1. MG/L medium: Add 5 g mannitol, 1 g L-glutamic acid, 250 mg KH_2PO_4, 100 mg NaCl, 100 mg $MgSO_4{\cdot}7H_2O$, 5 g tryptone, 2.5 g yeast extract, and 1 μg biotin to 900 mL distilled water. Adjust the pH to 7.0. Add distilled water to 1000 mL, filter sterilize using a 0.22-μm filter, and store at 4 °C.
2. Inoculum medium (WMS-inf): WMS-liq plus 100 μM acetosyringone (AS).
3. Cocultivation medium (WMS-AS): Add 0.46 g MS basic salts, 0.1 mL of 1000× modified MS vitamin stock, 10 g glucose, 0.5 g MES, and 1.25 mg $CuSO_4{\cdot}5H_2O$ to 900 mL of distilled water and adjust the pH to 5.8. Add distilled water to 1000 mL, add 8 g agarose (type I), and autoclave at 121 °C for 20 min. Cool to 50 °C and add 2 mL of 100 mM AS and 50 μL of 100 mM $AgNO_3$. Pour 20 mL of medium into each 90-mm Petri dish. Store at 4 °C in the dark.
4. Basic medium for callus culture (WMS): Add 4.6 g MS basic salts, 1 mL of 1000× modified MS vitamin stock, 0.5 g glutamine, 0.1 g casein hydrolysate, 0.75 g $MgCl_2{\cdot}6H_2O$, 40 g

maltose, and 1.95 g MES to 900 mL distilled water and adjust the pH to 5.8. Add 5 g agarose (type I) and autoclave at 121 °C for 20 min. Cool to 50 °C and add 0.5 mL of 1 mg/mL 2,4-D, 1 mL of 2.2 mg/mL picloram, 1 mL of 100 mg/mL ascorbic acid, and 50 μL of 100 mM $AgNO_3$. Pour 25 mL of medium into each 90-mm Petri dish. Store at RT in the dark.

5. Resting medium (WMS-Res): WMS plus 250 mg/L carbenicillin and 100 mg/L cefotaxime (added after autoclaving).
6. First selection medium (WMS-H15): WMS plus 250 mg/L carbenicillin and 15 mg/L hygromycin B (added after autoclaving).
7. Second selection medium (WMS-H30): WMS plus 250 mg/L carbenicillin and 30 mg/L hygromycin B (added after autoclaving).
8. Regeneration medium (WMSZ-H30): Add 4.6 g MS basic salts, 1 mL of 1000× modified MS vitamin stock, 20 g sucrose, 0.5 g MES, and 2.5 mg $CuSO_4{\cdot}5H_2O$ to 900 mL distilled water and adjust the pH to 5.8. Add distilled water to 1000 mL, add 3 g Phytagel, and autoclave at 121 °C for 20 min. Cool to 50 °C and add 1 mL of 5 mg/mL zeatin, 0.5 mL of 250 mg/mL carbenicillin, and 0.6 mL of 50 mg/mL hygromycin B. Pour 20 mL of medium into each 90-mm Petri dish. Store at RT in the dark.
9. Rooting medium (WMSF-H15): Add 4.6 g MS basic salts, 1 mL of 1000× modified MS vitamin stock, 15 g sucrose, 0.5 g MES, and 2.5 mg $CuSO_4{\cdot}5H_2O$ to 900 mL distilled water and adjust the pH to 5.8. Add distilled water to 1000 mL, add 3 g Phytagel, and autoclave at 121 °C for 20 min. Cool to 50 °C and add 0.3 mL of 50 mg/mL hygromycin. Pour 20 mL of medium into each 90-mm Petri dish. Store at RT in the dark.
10. The *hpt* primer set (forward 5′-GTGTCACGTTGCAAGACCTG-3′ and reverse 5′-GATGTTGGCGACCTCGTATT-3′) and the *Cas9* primer set (forward 5′-TATCACCATGCCCATGACGCTTATC-3′ and reverse 5′-TTTCCACCTTGGCTACTACCAACAC-3′) for detection of the transgenes.

3 Methods

3.1 Vector Construction

1. Identify a 20-nucleotide target sequence in the gene of interest using a website for gRNA design, such as CRISPR-direct (http://crispr.dbcls.jp/) or WheatCrispr (https://crispr.bioinfo.nrc.ca/WheatCrispr/) (*see* **Note 1**).
2. Synthesize forward and reverse DNA oligonucleotides.

3. Digest 2 μg of the pU6gRNA vector using 1 μL of *Bbs*I and 2 μL of CutSmart buffer in a 20-μL final volume and incubate at 37 °C for 3 h.
4. Run the digested products on a 1% agarose gel and purify the digested vector and gRNA expression cassette using a QIAquick Gel Extraction Kit.
5. Combine 1 μL of each oligo DNA (100 μM) and 48 μL sterile distilled water. Boil for 5 min and incubate at RT for 20 min to anneal the oligo.
6. Ligate the annealed gRNA oligo with the *Bbs*I-digested vector using a Rapid DNA Ligation Kit, combining 4 μL of digested vector (10 ng/μL), 4 μL of annealed oligo, 2 μL of 5× DNA dilution buffer, 10 μL of 2× T4 DNA ligation buffer, and 1 μL of T4 DNA ligase. Incubate at RT for 30 min.
7. Transform competent *E. coli* cells with 5 μL of the ligation reaction using an *E. coli* DH5α Competent Cells Kit. Plate the cells onto LB agar containing 50 mg/L ampicillin. Incubate the plate at 37 °C for 16 h.
8. Inoculate four colonies into separate 3-mL aliquots of liquid LB containing 50 mg/L ampicillin and grow the cells at 37 °C with vigorous shaking for 16 h.
9. Extract plasmid DNA with a QIAprep Miniprep Kit.
10. Sequence the plasmids with the OsU6-2F primer using a BigDye Terminator v3.1 Cycle Sequencing Kit to verify that the clones harbor the correct gRNA sequence (*see* **Note 2**).
11. Digest 2 μg of the binary vector pZH_ OsU6gRNA _PubiMMCas9 and the complete gRNA vector using 1 μL each of *Asc*I and *Pac*I and 2 μL of CutSmart buffer in a 20-μL final volume and incubate at 37 °C for 3 h.
12. Run the digested products on a 1% agarose gel and purify the digested vector and gRNA expression cassette using a QIAquick Gel Extraction Kit.
13. Ligate the digested gRNA expression cassette into the digested binary vector using a Rapid DNA Ligation Kit, combining 4 μL of digested vector (120 ng), 4 μL of gRNA cassette (10 ng), 2 μL of 5× DNA dilution buffer, 10 μL of 2× T4 DNA ligation buffer, and 1 μL of T4 DNA ligase. Incubate at RT for 30 min.
14. Transform competent *E. coli* cells with 5 μL of the ligation reaction. Plate the cells onto LB agar containing 75 mg/L spectinomycin and incubate overnight at 37 °C.
15. Inoculate four colonies into separate 5-mL aliquots of liquid LB containing 75 mg/L spectinomycin and grow the cells overnight at 37 °C with vigorous shaking.
16. Extract plasmid DNA with a QIAprep Miniprep Kit.

17. Sequence the plasmids with the OsU6-2F primer to verify that the clones harbor the correct gRNA sequence.
18. Inoculate *A. tumefaciens* strain EHA101 into 10 mL LB medium containing 50 mg/L kanamycin, and grow the cells overnight at 28 °C with vigorous shaking.
19. Collect bacterial cells by centrifugation and resuspend in 1 mL of precooled (on ice) 20 mM $CaCl_2$. Freeze 100 μL aliquots in liquid nitrogen.
20. Add 0.5–1.0 μg of plasmid (binary vector) DNA to the frozen cells directly and thaw for 5 min at 37 °C. Dilute in 1 mL of LB medium and incubate for 2 h at 28 °C with shaking. Plate the cells onto LB agar containing 75 mg/L spectinomycin, and incubate for 2 days at 28 °C.

3.2 Growing Wheat Plants to Obtain Immature Embryos

1. Sow 6 seeds of wheat cv. Fielder in 18-cm plastic pots containing a 2:1 mixture of Sakata Supermix A (fine peat moss) and Nippi fertilized granulated soil. In each pot, use 2.6 L of soil mixture and 2.6 g of controlled-release fertilizer (Osmocote Exact Mini 3-4 M 15-9-11 + 2MgO + TE).
2. Grow the plants in a glasshouse (Fig. 1a) with day/night temperatures of 16 °C/10 °C under an 8 h light/16 h dark photoperiod.
3. After 10–12 weeks (depending on the season), just before the heading stage, transfer the plants to a controlled environmental chamber (Fig. 1b) with day/night temperatures of 20 °C/13 °C under a 14-h light (300–500 μmol/m^2/sec)/10-h dark photoperiod.

3.3 Transient Assay to Evaluate the gRNA

1. Collect immature seeds at the correct developmental stage from panicles approximately 15 days after anthesis.

Fig. 1 Growing wheat plants to obtain immature embryos. Wheat plants growing in (**a**) a glasshouse and (**b**) a controlled environmental chamber

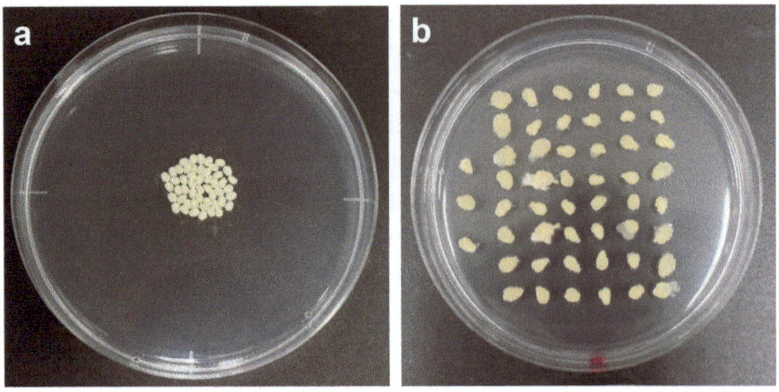

Fig. 2 Transient assay to evaluate gRNA. (**a**) Immature embryos before bombardment. (**b**) Callus tissue derived from embryos at 1 week after bombardment

2. Sterilize the immature seeds with 70% ethanol for 1 min and 0.5% sodium hypochlorite for 15 min and wash three times with sterile distilled water.
3. Isolate immature embryos from the immature seeds under a stereoscopic microscope and transfer the embryos into 2.0 mL of WMS-liq medium in a 2.0-mL microcentrifuge tube.
4. Centrifuge the microcentrifuge tube at 20,000× g at 4 °C for 10 min.
5. Transfer the immature embryos to WMS-Osm medium. Arrange ~50 embryos in a 2.0-cm-diameter circle in the center of each plate (Fig. 2a). Incubate the plate at 25 °C for at least 1 h before bombardment.
6. Mix 5 μg of each plasmid (i.e., one containing the gRNA and one harboring Cas9). Precipitate the mixed plasmid DNA in a mixture consisting of 1 mg gold particles (0.6 μm in diameter), 1 M $CaCl_2$, and 16 mM spermidine. Wash the gold particles twice with ethanol and resuspend in 50 μL ethanol.
7. Spread 15 μL of the gold particle suspension onto the central region of the macrocarrier of the PDS1000/He particle bombardment system and air dry in a laminar-flow clean bench.
8. After drying, perform bombardment according to the manufacturer's instructions, using 900 psi helium pressure and a target distance of 8 cm from the stopping screen.
9. Incubate bombarded immature embryos at 25 °C in the dark for 16 h. Transfer the embryos to WMS-CI medium and culture under the same conditions for 1 week (Fig. 2b).
10. Collect callus tissue derived from one embryo into a 2.0-mL microcentrifuge tube and add 300 μL of DNA extraction buffer.

11. Homogenize the sample using a TissueLyser II with zirconium beads. After homogenizing, add 150 μL of 10% SDS and incubate at RT for 15 min.
12. Centrifuge at 2700× g at RT for 10 min.
13. Transfer 100 μL of the supernatant to a new centrifuge tube and add 10 μL of 3 M sodium acetate and 100 μL isopropanol. Mix thoroughly and centrifuge at 15,000× *g* at RT for 10 min.
14. Discard the supernatant and wash the pellet with 70% ethanol. Remove the ethanol and air-dry completely. Dissolve the DNA pellet in 0.1× TE buffer.
15. PCR-amplify the fragment including the target site.
16. Subject half of the PCR product to agarose gel electrophoresis to confirm the presence of a single band.
17. Digest the PCR product with the appropriate restriction enzyme according to the manufacturer's instructions. Subject the digested PCR product to agarose gel electrophoresis to detect the presence of the mutation as an undigested band (*see* **Note 3**).
18. Extract the PCR product containing the mutation from the agarose gel. Purify the PCR product, clone it in a cloning vector such as pCRII from the TA Cloning Kit, and transform it into *E. coli*. Extract the plasmid and subject it to Sanger sequencing.

3.4 Agrobacterium-Mediated Transformation

1. Culture *A. tumefaciens* in 10 mL of MG/L medium (no antibiotics) overnight at 28 °C with vigorous shaking. Collect bacterial cells by centrifugation and resuspend in the same volume of WMS-inf medium.
2. Isolate immature embryos (*see* **Note 4**) and transfer into 2-mL microcentrifuge tubes as described above.
3. Invert each microcentrifuge tube several times and remove the medium. Add 2.0 mL WMS-liq medium. Centrifuge the tube at 20,000× *g* at 4 °C for 10 min.
4. Remove the medium from the microcentrifuge tube and add 1.0 mL inoculum. Invert the microcentrifuge tube repeatedly for 30 s and incubate at RT for 5 min.
5. Transfer the immature embryos to WMS-AS medium (Fig. 3a). Gently turn over the immature embryos and place on the medium scutellum side up. Seal the plate with Parafilm and incubate at 23 °C in the dark for 2 days.
6. After 2 days of cocultivation, excise the embryo axis from each immature embryo using a scalpel and forceps. Transfer the immature embryos to WMS-Res medium scutellum side up

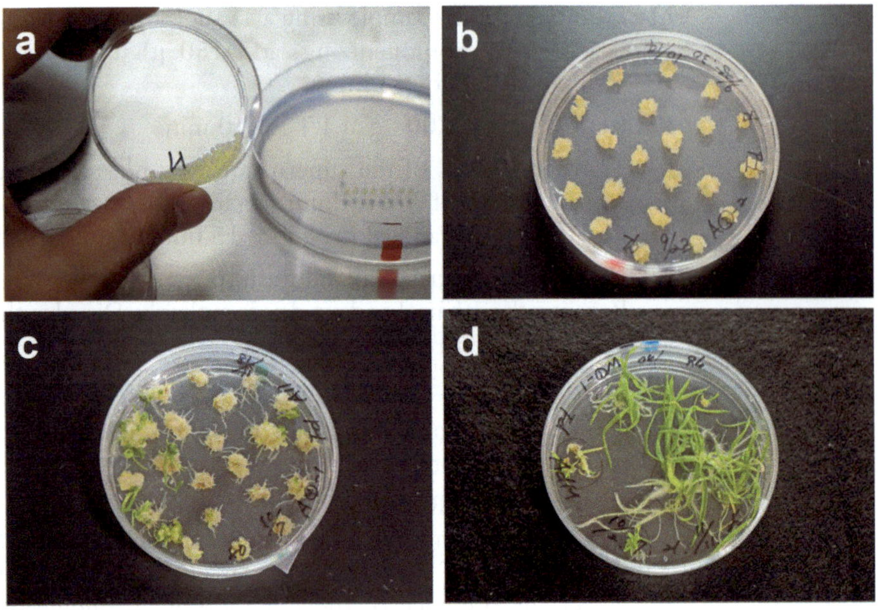

Fig. 3 *Agrobacterium*-mediated transformation. (**a**) Infection of immature wheat embryos with *A. tumefaciens.* (**b**) Callus tissues after 3 weeks of culture on second selection medium. (**c**) Regenerated tissues after 2 weeks of culture on regeneration medium. (**d**) Rooted wheat plant

for resting culture. Seal the plate with micropore surgical tape and culture at 25 °C in the dark for 5 days.

7. After 5 days of resting culture, transfer the immature embryos to WMS-H15 medium for first selection and culture at 25 °C in the dark for 2 weeks.
8. Transfer the resulting calli to WMS-H30 medium for second selection and culture under the same conditions for 3 weeks (Fig. 3b).
9. After selection, transfer the calli to WMSZ-H30 medium for regeneration. Seal the plate with Parafilm and culture at 25 °C under a 14-h light (68 μmol/m^2/sec)/10-h dark photoperiod for 2 weeks (Fig. 3c).
10. Transfer regenerated plants to WMSF-H15 medium for rooting and culture under the same conditions for 2 weeks (Fig. 3d). Remove the Parafilm, seal the plates with Micropore surgical tape, and culture for 3–5 more days (*see* **Note 5**).
11. Transplant rooted plants to soil.
12. Extract DNA from the leaves of the plants as described above. Confirm the presence of the transgene by PCR with the *hpt* primer set and the *Cas9* primer set.
13. Detect the mutation in the transformant as described above.

4 Notes

1. To produce a genome-edited triple recessive mutant in wheat, the target sequences must be present in all homoeologs, i.e., the A, B, and D sub-genomes, and there must be a 20-nucleotide region preceding a PAM (5′-NGG-3′) that is recognized by Cas9.
2. It is better if the candidate target site has a restriction enzyme site that overlaps with the fourth nucleotide from the PAM; this makes it easier to detect mutations by PCR-RFLP using that restriction enzyme. If such a site is not available, the PCR products, including the target sites of all homoeologs, should be sequenced to detect the mutations using a technique such as amplicon sequencing.
3. The plasmid including the gRNA expression cassette can be used directly to evaluate the gRNA.
4. It is critical to use immature embryos at the correct developmental stage. The size of an embryo is a very good indicator of its stage: Immature embryos 2.0–2.5 mm in length along the long axis are optimal for transformation.
5. Changing from Parafilm to Micropore surgical tape 3–5 days before transplanting helps the rooted plants acclimate to the new growth environment.

References

1. Lu Y, Ye X, Guo R, Huang J, Wang W, Tang J, Tan L, Zhu J, Chu C, Qian Y (2017) Genome-wide targeted mutagenesis in rice using the CRISPR/Cas9 system. Mol Plant 10:1242–1245
2. Zhang Y, Massel K, Godwin ID, Gao C (2018) Applications and potential of genome editing in crop improvement. Genome Biol 19:210
3. Wang Y, Cheng X, Shan Q, Zhang Y, Liu J, Gao C, Qiu JL (2014) Simultaneous editing of three homoeoalleles in hexaploid bread wheat confers heritable resistance to powdery mildew. Nat Biotechnol 32:947–951
4. Wang K, Gong Q, Ye X (2019) Recent developments and applications of genetic transformation and genome editing technologies in wheat. Theor Appl Genet 133:1603–1622
5. Abe F, Haque E, Hisano H, Tanaka T, Kamiya Y, Mikami M, Kawaura K, Endo M, Onishi K, Hayashi T, Sato K (2019) Genome-edited triple-recessive mutation alters seed dormancy in wheat. Cell Rep 28:1362–1369
6. Arndell T, Sharma N, Langridge P, Baumann U, Watson-Haigh NS, Whitford R (2019) gRNA validation for wheat genome editing with the CRISPR-Cas9 system. BMC Biotechnol 19:71
7. Brandt KM, Gunn H, Moretti N, Zemetra RS (2020) A streamlined protocol for wheat (*Triticum aestivum*) protoplast isolation and transformation with CRISPR-Cas ribonucleoprotein complexes. Front Plant Sci 11:769
8. Kamiya Y, Abe F, Mikami M, Endo M, Kawaura K (2020) A rapid method for detection of mutations induced by CRISPR/Cas9-based genome editing in common wheat. Plant Biotechnol 37:247–251
9. Mikami M, Toki S, Endo M (2015) Comparison of CRISPR/Cas9 expression constructs for

efficient targeted mutagenesis in rice. Plant Mol Biol 88:561–572

10. Hood EE, Helmer GL, Fraley RT, Chilton MD (1986) The hypervirulence of *Agrobacterium tumefaciens* A281 is encoded in a region of pTiBo542 outside of T-DNA. J Bacteriol 168: 1291–1301

11. Hamada H, Liu Y, Nagira Y, Miki R, Taoka N, Imai R (2018) Biolistic-delivery-based transient CRISPR/Cas9 expression enables *in planta* genome editing in wheat. Sci Rep 8: 14422

Chapter 14

Targeted Modification of Grain Dormancy Genes in Barley

Hiroshi Hisano, Robert E. Hoffie, Jochen Kumlehn, and Kazuhiro Sato

Abstract

Transgenesis technologies, such as overexpression or RNA interference-mediated suppression, have often been used to alter the activity of target genes. More recently developed targeted genome modification methods using customizable endonucleases allow for the regulation or knockout mutation of target genes without the necessity of integrating recombinant DNA. Such approaches make it possible to create novel alleles of target genes, thereby significantly contributing to crop improvement. Among these technologies, the Cas9 endonuclease-based method is widely applied to several crops, including barley (*Hordeum vulgare*). In this chapter, we describe an *Agrobacterium*-based approach to the targeted modification of grain dormancy genes in barley using RNA-guided Cas9 nuclease.

Key words *Agrobacterium tumefaciens*, Cas9 nuclease, CRISPR, Genetic engineering, *Hordeum vulgare*, Preharvest sprouting

1 Introduction

Grain dormancy is a quantitative trait that is controlled by multiple genes. Barley varieties with short dormancy are more suitable for malting, but dormancy that is too short can cause preharvest sprouting. To control the balance between dormancy and germination, it is important to find new alleles of dormancy-related genes for barley breeding. Quantitative trait loci (QTL) associated with grain dormancy in barley have been identified using several genetic mapping populations, and their responsible genes have been isolated [1, 2]. Natural variants or artificially induced mutants also provide a source of new alleles. However, isolating the desired alleles from these genetic resources is costly and laborious. Alternatively, the expression of dormancy-related genes can be modulated via transgenesis such as overexpression and suppression, representing a more efficient strategy than conventional breeding techniques.

Naoto Kawakami and Kazuhiro Sato (eds.), *Seed Dormancy: Methods and Protocols*, Methods in Molecular Biology, vol. 2830,
https://doi.org/10.1007/978-1-0716-3965-8_14,

Targeted genome modification systems using customizable endonucleases have recently been developed as efficient means for generating novel alleles or regulating target genes. These methods are collectively referred to as "genome editing." Customizable endonucleases trigger double-strand breaks in the target genome sequence, thus functioning as molecular scissors. Deletions or insertions occur during subsequent DNA repair. Various platforms of customizable endonucleases, namely, zinc finger nucleases (ZFNs), transcription activator-like effector nucleases (TALENs), and clustered regularly interspaced short palindromic repeats (CRISPR)-associated protein 9 (Cas9), have been utilized for targeted genome modification in barley. Among these, the Cas9-based method is the most widely used due to its simplicity and versatility. Lawrenson et al. [3] were the first to apply Cas9 endonuclease to barley and confirmed the insertion of a mutation in the *HvPM19* gene, which encodes an abscisic acid-inducible plasma membrane protein involved in grain dormancy. Mutant barley lines have since been created via Cas9-mediated genome editing for several agronomically important genes, such as the *NUD1* gene involved in determining hulled/hulless grains [4].

CRISPR/Cas9 was discovered as an immune system that protects microorganisms against invading viruses and plasmids. In its original configuration, CRISPR RNA (crRNA) and transactivated CRISPR RNA (tracrRNA) form a complex that directs Cas9 nuclease to the foreign DNAs. To facilitate genome editing, these crRNAs and tracrRNAs were integrated into a single guide RNA (gRNA) that can be readily customized to address virtually any genomic target motif of choice [5]. To perform targeted genome modification, it is necessary to introduce the components of customizable nucleases into cells of the host plant, which is mostly conducted via transgenesis. SpCas9 from *Streptococcus pyogenes* is the most widely used among the several Cas nucleases isolated from various species. By customizing the 20 bases at the 5′ end of the gRNA according to a genomic target motif, Cas9 protein becomes bound to a 5′-NGG-3′ protospacer adjacent motif (PAM) that constitutes an additional part of the target motif and resides immediately downstream of the genomic sequence that matches up with the gRNA. In this system, Cas9 nuclease mediates the formation of a double-strand break approximately three to four bases upstream of the PAM.

We recently succeeded in obtaining genome-edited barley with mutations in the grain dormancy genes *QTL FOR SEED DORMANCY Qsd1* and *Qsd2* using *Agrobacterium tumefaciens* and RNA-guided Cas9 nuclease [6]. In this chapter, we explain how to obtain genome-modified mutant barley by the Cas9-based method via *Agrobacterium*-mediated transformation. We first describe the preparation of plant materials (Fig. 1), the design of the guide RNA, and vector construction. Then, *Agrobacterium*-

Fig. 1 Growing barley to obtain immature embryos. Barley plants grown in growth chambers under (**a**) a short-day photoperiod (12 h light at 15 °C and 12 h darkness at 13 °C) for the initial stage and (**b**) a long-day photoperiod (16 h light at 16 °C and 8 h darkness at 13 °C) for the later stage

meditated transformation using immature embryos of barley is described (Figs. 2, 3), followed by the sequencing analysis of target genes in regenerated plants (Fig. 4).

2 Materials

2.1 Plant Material

Immature embryos of barley (*Hordeum vulgare*) are used as the recipient explant for DNA transfer, since they contain highly totipotent cells that give rise to somatic embryos or adventitious shoots for plant regeneration. This investigation was conducted using the donor cv. "Golden Promise," as it constitutes the most suitable accession for genetic engineering in barley [7, 8].

2.2 Media for Bacterial Culture

1. LB medium: 10 g/L tryptone, 5 g/L yeast extract, and 0.5 g/L NaCl (15 g/L agar for solid medium), adjusted to pH 7.0 and autoclaved. After autoclaving, add appropriate antibiotics.
2. MG/L medium: 0.25 g/L KH_2PO_4, 0.1 g/L NaCl, 0.1 g/L $MgSO_4$-$7H_2O$, 5 g/L mannitol, 5 g/L tryptone, 2.5 g/L yeast extract, and 1 g/L glutamic acid (15 g/L agar for solid medium), adjusted to pH 7.0 and autoclaved. After autoclaving, add 10 μL/L of 100 mg/L biotin solution.

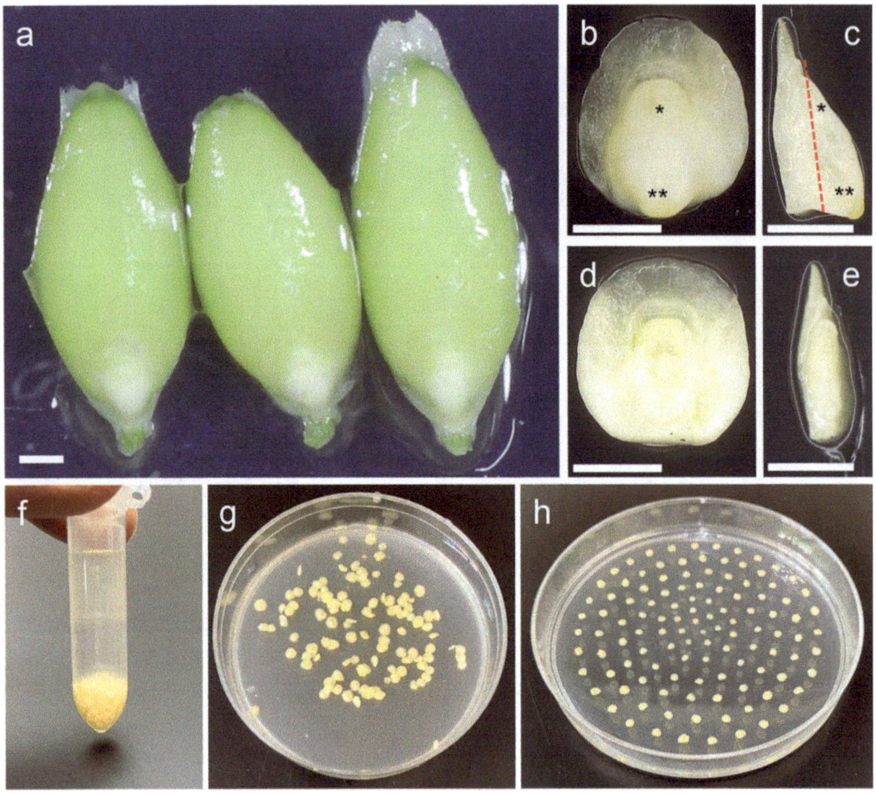

Fig. 2 *Agrobacterium*-mediated transformation of barley. (**a**) Immature barley grains without lemma. (**b–e**) Immature barley embryos containing a plumule (*) and primary root meristem (**) in top view (**b**), cross-section view (**c**), and with the plumule and root meristem removed in top (**d**) and cross-section (**e**) views. Red dotted line shows the cut position. (**f**) Immature embryos collected in 1.5 mL of autoclaved deionised water in a 2-mL microcentrifuge tube. (**g**) Immature embryos in *Agrobacterium* inoculum in a 6-cm Petri dish. (**h**) Coculture of immature embryos and *Agrobacterium* on CCM in a 9-cm Petri dish. Size bar indicates 1 mm (**a, b, c, d, e**)

Fig. 3 Selection of transformants. (**a**) Resistant callus after 4 weeks of culture on CSM. (**b**) Regenerated shoots after 2 weeks of culture on RGM. (**c**) Rooted barley plant on ROM. Size bars indicate 3 mm (**a, b**) and 10 mm (**c**)

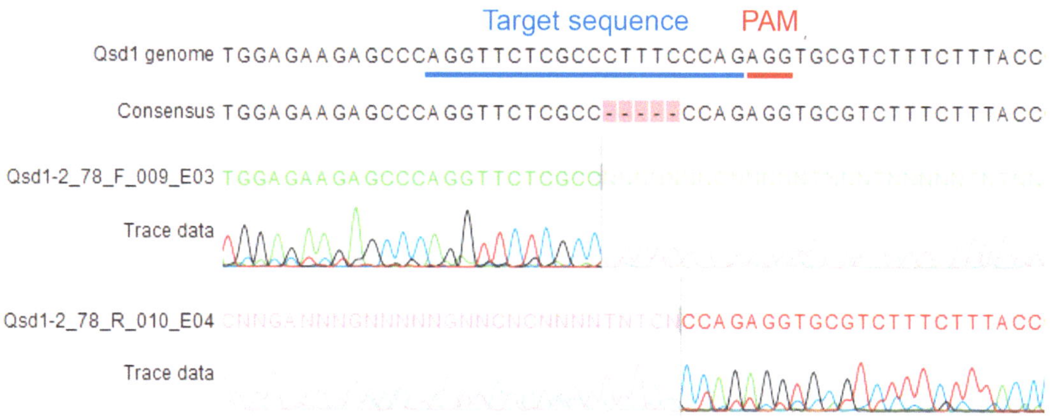

Fig. 4 Sequencing of the target gene. Example of Sanger sequencing results to confirm the introduction of a mutation. The gRNA target sequence and the PAM sequence are underlined in blue and red, respectively. In the case of a heterozygous mutation, two different sequences that make it unreadable appear around the mutation site, as shown in the figure. In this example, five bases (CTTTC) have been deleted

2.3 Media for Barley Transformation

See Table 1 for the compositions of the media listed below:

1. Liquid infection medium (LIM): Add 100 mL of 10× LS major mineral stock, 10 mL of 100× LS minor mineral stock, 10 mL of 100× NA_2-EDTA-Fe stock, 1 g casamino acids, 0.69 g L-proline, 0.35 g myo-inositol, 1 mg thiamine HCl, 2.5 mg 3,6-dichloro-2-methoxybenzoic acid (Dicamba), and 30 g maltose to 700 mL of MilliQ water and adjust the pH to 5.8. Adjust the volume to 1000 mL with MilliQ water. Autoclave at 121 °C for 20 min and cool to 50 °C. Add 19.6 mg acetosyringone. Store at 4 °C in the dark.
2. Coculture medium (CCM): Add 100 mL of 10× LS major mineral stock, 10 mL of 100× LS minor mineral stock, 10 mL of 100× NA_2-EDTA-Fe stock, 1 g casamino acids, 0.69 g L-proline, 0.35 g myo-inositol, 1 mg thiamine HCl, 2.5 mg Dicamba, and 30 g maltose to 700 mL of MilliQ water and adjust the pH to 5.8. Adjust the volume to 1000 mL with MilliQ water and add 3.5 g Gelrite. Autoclave at 121 °C for 20 min and cool to 50 °C. Add 19.6 mg acetosyringone. Pour 20 mL of medium into each 90-mm Petri dish (15-mm depth). Store at 4 °C in the dark.
3. Callus induction (resting) medium (CIM): Add 100 mL of 10× LS major mineral stock, 10 mL of 100× LS minor mineral stock, 10 mL of 100× NA_2-EDTA-Fe stock, 1 g casamino acids, 0.69 g L-proline, 0.35 g myo-inositol, 1 mg thiamine HCl, 2.5 mg Dicamba, and 30 g maltose to 700 mL of MilliQ water and adjust the pH to 5.8. Adjust the volume to 1000 mL with MilliQ water and add 3.5 g Gelrite. Autoclave at 121 °C for 20 min and cool to 50 °C. Add 1.95 mg $CuSO_4{\cdot}5H_2O$,

Table 1
Composition of the media used for barley tissue culture

	LIM	CCM	CIM	CSM	RGM	ROM
LS major mineral stock (×10)	100 mL/L	100 mL/L	100 mL/L	100 mL/L		50 mL/L
LS minor mineral stock (×100)	10 mL/L	10 mL/L	10 mL/L	10 mL/L		10 mL/L
K4N major mineral stock (×10) [12]					100 mL/L	
K4N minor mineral stock (×100) [12]					10 mL/L	
K4N vitamin stock (×100) [12]					10 mL/L	
NA_2-EDTA-Fe stock (×100)	10 mL/L	10 mL/L	10 mL/L	10 mL/L	10 mL/L	10 mL/L
MS vitamin stock (×100)						10 mL/L
Casamino acids	1 g/L	1 g/L	1 g/L	1 g/L		0.5 g/L
L-proline	0.69 g/L	0.69 g/L	0.69 g/L	0.69 g/L		
Myo-inositol	0.35 g/L	0.35 g/L	0.35 g/L	0.35 g/L		
Thiamine HCl	1 mg/L	1 mg/L	1 mg/L	1 mg/L		
Dicamba	2.5 mg/L	2.5 mg/L	2.5 mg/L	2.5 mg/L		
6-Benzylaminopurine					0.225 mg /L	
Indole-3-butyric acid						0.2 mg/L
Maltose	30 g/L	30 g/L	30 g/L	30 g/L	36 g/L	
Sucrose						15 g/L
Gelrite		3.5 g/L	3.5 g/L	3.5 g/L	4 g/L	3 g/L
pH	5.8	5.8	5.8	5.8	5.8	5.8
*Add after autoclaving						
*Acetosyringone	19.6 mg/L	19.6 mg/L				
*$CuSO_4 \cdot 5H_2O$			1.95 mg/L	1.95 mg/L		
*Timentin			250 mg/L	250 mg/L	150 mg/L	
*Cefotaxime			250 mg/L	250 mg/L	150 mg/L	
*Hygromycin				75 mg/L	30 mg/L	30 mg/L

250 mg timentin, and 250 mg cefotaxime. Pour 20 mL of medium into each 90-mm Petri dish (15-mm depth). Store at 4 °C in the dark.

4. Callus selection medium (CSM): Add 100 mL of 10× LS major mineral stock, 10 mL of 100× LS minor mineral stock, 10 mL of 100× NA_2-EDTA-Fe stock, 1 g casamino acids, 0.69 g L-proline, 0.35 g myo-inositol, 1 mg thiamine HCl, 2.5 mg Dicamba, and 30 g maltose to 700 mL of MilliQ water and adjust the pH to 5.8. Adjust the volume to 1000 mL with MilliQ water and add 3.5 g Gelrite. Autoclave at 121 °C for 20 min and cool to 50 °C. Add 1.95 mg $CuSO_4{\cdot}5H_2O$, 250 mg timentin, 250 mg cefotaxime, and 75 mg hygromycin. Pour 20 mL of medium into each 90-mm Petri dish (15-mm depth). Store at 4 °C in the dark.
5. Regeneration medium (RGM): Add 100 mL of 10× K4N major mineral stock, 10 mL of 100× K4N minor mineral stock, 10 mL of 100× K4N vitamin stock, 10 mL of 100× NA_2-EDTA-Fe stock, 0.225 mg 6-benzylaminopurine, and 36 g maltose to 700 mL of MilliQ water and adjust the pH to 5.8. Adjust the volume to 1000 mL with MilliQ water and add 4 g Gelrite. Autoclave at 121 °C for 20 min and cool to 50 °C. Add 150 mg timentin, 150 mg cefotaxime, and 30 mg hygromycin. Pour 25 mL of medium into each 90-mm Petri dish (20-mm depth). Store at 4 °C in the dark.
6. Rooting medium (ROM): Add 50 mL of 10× LS major mineral stock, 10 mL of 100× LS minor mineral stock, 10 mL of 100× NA2-EDTA-Fe stock, 10 mL of 100× MS vitamin stock, 0.5 g casamino acids, 0.2 mg indole-3-butyric acid, and 15 g sucrose to 800 mL of MilliQ water and adjust the pH to 5.8. Adjust the volume to 1000 mL with MilliQ water and add 3 g Gelrite. Autoclave at 121 °C for 20 min and cool to 50 °C. Add 30 mg hygromycin. Pour 40 mL of medium into each plant culture box (75 × 75 × 100 mm). Store at 4 °C in the dark.

2.4 Vector Construction and Gene Analysis

Agrobacterium-mediated transformation is employed to deliver expression units for gRNA and *cas9* into the barley genome.

1. Vectors: p6i-2×35S-TE9 and pSH121 [9] (*see* **Note 1**).
2. Restriction enzymes: *Bsa*I, *Sfi*I, and *Eco*RI.
3. Ligase: T4 DNA ligase (e.g., Quick Ligation Kit, New England Biolabs).
4. Competent *Escherichia coli* cells.
5. Incubator with a shaker at constant temperature.
6. Taq DNA polymerase.

7. Plasmid extraction kit (e.g., GeneJET Plasmid Miniprep Kit, Thermo Fisher Scientific).
8. Gel electrophoresis system.
9. Gel extraction kit (e.g., GeneJET PCR Purification Kit, Thermo Fisher Scientific).
10. *Agrobacterium tumefaciens* strain AGL1 or EHA105.
11. DNA extraction kit (e.g., DNeasy Plant mini kit, QIAGEN, Easy DNA Extraction Kit version 2, KANEKA, Japan).

3 Methods

3.1 Prepare Plant Materials

1. Initial growth: Grow five plants in a 3.5-L unglazed pot in commercially available soil comprising peat moss, bark compost, leaf mold, and sand at a 4:3:1:2 ratio. Plants are initially grown in a growth chamber for 8 weeks under short-day conditions at 15/13 °C (day/night) with 12 h of daylight (Fig. 1a).
2. Harvesting immature embryos: Transfer plants to long-day conditions at 16/13 °C with 16 h of daylight (Fig. 1b). For "Golden Promise," heading begins approximately 3–4 weeks after moving to long-day conditions, and immature embryos can be collected ~2 weeks after pollination.

3.2 Selection of Target Motifs and Design of Guide RNAs

1. Use the genomic DNA sequences of the target genes of the plant genotype to be modified.
2. Online tools such as WU-CRISPR (http://crisprdb.org/wu-crispr-website/ [10]) are instrumental for the preselection of suitable target motifs.
3. Check the positions of the target motifs: Choose those with the cleavage site residing in the exons, including a promoter-compatible transcription start and with a platform score as high as possible (*see* **Note 2**).
4. Selection of gRNAs: Generate the predicted secondary structures of the gRNAs using the RNAfold online platform (http://rna.tbi.univie.ac.at//cgi-bin/RNAWebSuite/RNAfold.cgi), and select the best targets following the criteria described by Kumlehn et al. [11].

3.3 Vector Construction

1. Order DNA oligonucleotides like standard primers, including the identified 20-nucleotide target sequence forward and reverse, respectively, and compatible overhangs with the cloning vector. For example, using pSH121 as the cloning vector and gRNA2 of *Qsd1* (5′-AGGTTCTCGCCCTTTCCCAG-3′) as the target sequence, the following oligos were used:

Qsd1_gRNA2_f (5′-tggcAGGTTCTCGCCCTTTCCCAG-3′) and Qsd1_gRNA2_r (5′-aaac CTGGGAAAGGGCGA GAACCT-3′).

2. Dilute each oligonucleotide to 10 μM concentration. Mix 5 μL of both diluted oligonucleotides and heat up to 98 °C for 5 min followed by a slow temperature decrease of 0.2 °C/s to room temperature (ca. 23 °C) in a thermocycler.
3. Using Quick Ligation Kit (0.5 μL of enzyme and 5 μL of 2× buffer), ligate 1.5 μL of annealed oligonucleotides and 70 ng of pSH121, predigested by the restriction enzyme *Bsa*I in a 10-μL reaction mixture.
4. Transform 4 μL of the ligation reaction mixture into 50 μL of *E. coli* competent cells. Plate the cells onto LB agar containing 50 mg/L kanamycin and incubate at 37 °C overnight.
5. Pick *E. coli* colonies with a sterilized toothpick and dip them shortly into 10 μL PCR reaction mixture containing 0.5 μM specific primers, pSH121-gRNAcheckF1 (5′- GGAAAC CACGTGATGTGAAG -3′), and Qsd1_gRNA2_r for *Qsd1*, followed by dipping the toothpick into 100 μL LB medium containing 50 mg/L kanamycin.
6. Perform PCR using the following program: initial denaturation at 95 °C for 10 min, 35 cycles of denaturation at 95 °C for 30 s, annealing at 55 °C for 30 s, and extension at 72 °C for 30 s, with a final extension at 72 °C for 10 min. Check them by gel electrophoresis.
7. Inoculate each PCR-positive colony into 4 mL of liquid LB containing 50 mg/L kanamycin and grow the cells at 37 °C with shaking (ca. 120 rpm) overnight. Extract the plasmid DNA from each colony with a plasmid extraction kit.
8. Sanger-sequence the plasmids with the pSH121-gRNAcheckF1 primer to verify that the clones harbor the correct 20-nucleotide target sequence.
9. Digest 200 ng of the destination binary vector p6i-2×35S-TE9 and pSH121 including gRNA with the target sequence and *cas9* expression cassettes using 2 μL *Sfi*I and 4 μL of buffer G in a 40-μL final volume and incubate at 50 °C for 3 h.
10. Run the digested products on a 1% agarose gel with TBE buffer and excise the DNA fragments of 7600 bp with the *cas9* and gRNA expression units and 10,000 bp from p6i-2×35S-TE9. Extract and purify each DNA fragment using a gel extraction kit.
11. Ligate the extracted *cas9* and gRNA expression cassettes (50 ng) and the digested p6i-2×35S-TE9 vector (20 ng) using a Quick Ligation Kit (0.5 μL of enzyme and 5 μL of

2× buffer) in a total volume of 10 μL. Incubate at room temperature for 20 min.

12. Transform 5 μL of the ligation reaction mixture into 50 μL of *E. coli* competent cells. Plate the cells onto LB agar containing 100 mg/L spectinomycin and incubate overnight at 37 °C.
13. Pick four colonies with a sterile toothpick into a 4-mL aliquot of liquid LB containing 100 mg/L spectinomycin and grow the cells at 37 °C with shaking (ca. 120 rpm) overnight.
14. Extract plasmid DNA with a plasmid extraction kit and perform a test digest using 2 μL 5× *Eco*RI-Buffer, 0.5 μL *Eco*RI restriction enzyme, and 0.75 μL plasmid DNA in a total reaction volume of 10 μL and incubate at 37 °C for 2 h. Run a 1% agarose gel electrophoresis to confirm the band sizes of 8500 bp, 5200 bp, and 1400 bp.
15. Dilute the extracted plasmid (50 ng) of a confirmed clone 1:10 in water and use 1 μL of the dilution for electroporation of 50 μL of *Agrobacterium* strain AGL1- or EHA105-competent cells.
16. Plate the cells onto MG/L agar medium containing 100 mg/L spectinomycin, 100 mg/L carbenicillin, and 50 mg/L rifampicin and incubate at 28 °C for 72 h.

3.4 Agrobacterium-Mediated Transformation

1. Culture *Agrobacterium* in 10 mL of MG/L medium with 50 mg/L streptomycin in a 50-mL conical tube for 18 h at 28 °C with shaking (ca. 80 rpm).
2. Collect the bacterial cells by centrifugation (1972 × *g*, 10 min, 20 °C) and resuspend in LIM, adjusting the $O.D._{660}$ to 0.60–0.65.
3. Harvest grains containing ca. 1.5–2.0-mm-long immature embryos (*see* **Note 3**).
4. Remove the lemma using tweezers and sterilize the grains with 70% ethanol for 1 min and sodium hypochlorite (1% chloride concentration) with 0.1% Tween20 for 5–10 min.
5. Rinse three times with autoclaved deionised water in a clean bench (Fig. 2a-c).
6. Excise the plumule and radicle with a scalpel and isolate the immature embryo from each immature grain under a stereomicroscope (Fig. 2d, e) (*see* **Note 4**). Place the immature embryo on solid medium (autoclaved 0.8% agar and water) in a 6-cm Petri dish.
7. Place the immature embryos in a 2-mL microcentrifuge tube containing 1.5 mL autoclaved deionised water plus 19.6 mg/L acetosyringone (Fig. 2f).

8. Remove the autoclaved deionised water and add 1.5 mL fresh autoclaved deionised water containing 19.6 mg/L acetosyringone.
9. Incubate the immature embryos at 43 °C for 4 min (*see* **Note 5**) and cool on ice for 2 min. Remove the water and add 1 mL of *Agrobacterium* inoculum to the tube.
10. Decant the inoculum with the immature embryos into an empty 6-cm Petri dish and add the remaining inoculum (Fig. 2g).
11. Incubate for 5 min at room temperature (ca. 25 °C).
12. Discard the inoculum and rinse once with LIM. Place the immature embryos on CCM scutellum side down (Fig. 2h).
13. Seal the plate with micropore surgical tape and co-cultivate at 25 °C in the dark for 3 days.
14. After 3 days of cocultivation, transfer the immature embryos to CIM scutellum side down (*see* **Note 6**). Seal the plate with micropore surgical tape and culture at 25 °C in the dark for 7 days.
15. Transfer the resulting calli to CSM for the first selection and culture at 25 °C in the dark for 2 weeks.
16. Transfer the calli to fresh CSM for the second selection and subculture under the same conditions for 2 weeks.
17. Transfer the hygromycin-resistant calli (Fig. 3a) to RGM for regeneration. Seal the plate with micropore surgical tape and culture at 25 °C under a 16-h-light (53.0 μmol m^{-2} s^{-1} PPF in 400–700 nm)/10-h-dark photoperiod for 2 weeks.
18. Transfer the regenerated plants (Fig. 3b) to ROM for rooting and culture under the same conditions for 4 weeks or subculture them on RGM for 2 more weeks.

3.5 Screening and Transplanting Regenerated Plants

1. When the shoots and roots have fully grown on ROM (>5 cm and 2 cm, respectively) (Fig. 3c), harvest the shoots in a clean bench.
2. Extract DNA from the collected shoots with a DNA Extraction Kit (KANEKA Easy DNA Extraction Kit version 2).
3. PCR confirmation: Confirm the presence of the transgene by PCR with the *hpt* primer set (forward 5′-GCTGGGGCGTC GGTTTCCACTATCGG -3′ and reverse 5′- CGCATAA CAGCGGTCATTGACTGGAGC-3′) and the *cas9* primer set (forward 5′-AGAATCAGACCACGCAGAAG-3′ and reverse 5′-GGACCTTGTTGTCAATGGAG-3′).
4. Confirm the sequence of the target gene: Perform PCR using the extracted DNA described above and specific primers for the target region of the gene, e.g., the *Qsd1*_gRNA2 primer set

(forward 5′- CACGCAGAGGTAACAGCAAC -3′ and reverse 5′- AGAGCCTGAGGGTTCCCAAG -3′). Subject the PCR fragments to Sanger sequencing (Fig. 4).

5. Transfer plants with confirmed introduced mutations to plant culture boxes and incubate in an isolated greenhouse or growth chamber for 1–2 days with the lid open.
6. Transplant the plants to soil and grow under long-day conditions at 16/13 °C with 16 h of daylight.

4 Notes

1. pSH121 contains the rice-derived *OsU3* promoter for gRNA expression and a maize codon-optimized *SpCas9* driven by the maize *Ubiquitin* promoter.
2. SpCas9 cleaves between the third and fourth bases upstream of the PAM site; the location should be within an exon for gene knockout.
3. On the day of the *Agrobacterium* inoculation, sample one grain from each spike and check the size of the embryo; if a healthy, creamy-yellow, 1.5–2.0-mm-long embryo has formed, harvest the spikes. If the embryo is <1.5 mm long, it is likely to die following cocultivation with *Agrobacterium*, and if it is >2.0 mm long, it is unlikely that embryogenic callus will be obtained.
4. If it is cumbersome to remove the plumule, the removal of the root meristem is in fact sufficient to prevent germination.
5. Do not incubate the immature embryo at 43 °C for more than 4 min.
6. If *Agrobacterium* growth is observed at this point, wash explants with autoclaved deionised water. In addition, the O. $D._{600}$ of the *Agrobacterium* should be lower at the infection step.

References

1. Sato K, Yamane M, Yamaji N, Kanamori H, Tagiri A, Schwerdt JG, Fincher GB, Matsumoto T, Takeda K, Komatsuda T (2016) Alanine aminotransferase controls seed dormancy in barley. Nat Commun 7:11625
2. Nakamura S, Pourkheirandish M, Morishige H, Kubo Y, Nakamura M, Ichimura K, Seo S, Kanamori H, Wu J, Ando T, Hensel G, Sameri M, Stein N, Sato K, Matsumoto T, Yano M, Komatsuda T (2016) *Mitogen-activated protein kinase kinase 3* regulates seed dormancy in barley. Curr Biol 26:775–781
3. Lawrenson T, Shorinola O, Stacey N, Li C, Østergaard L, Patron N, Uauy C, Harwood W (2015) Induction of targeted, heritable mutations in barley and *Brassica oleracea* using RNA-guided Cas9 nuclease. Genome Biol 16:258
4. Gerasimova SV, Hertig C, Korotkova AM, Kolosovskaya EV, Otto I, Hiekel S, Kochetov AV, Khlestkina EK, Kumlehn J (2020)

Conversion of hulled into naked barley by Cas endonuclease-mediated knockout of the *NUD* gene. BMC Plant Biol 20:255

5. Jinek M, Chylinski K, Fonfara I, Hauer M, Doudna JA, Charpentier E (2012) A programmable dual-RNA–guided DNA endonuclease in adaptive bacterial immunity. Science 337: 816–821
6. Hisano H, Hoffie RE, Abe F, Munemori H, Matsuura T, Endo M, Mikami M, Nakamura S, Kumlehn J, Sato K (2022) Regulation of germination by targeted mutagenesis of grain dormancy genes in barley. Plant Biotechnol J 20: 37–46
7. Hisano H, Meints B, Moscou MJ, Cistue L, Echávarri B, Sato K, Hayes PM (2017) Selection of transformation-efficient barley genotypes based on *TFA* (*transformation amenability*) haplotype and higher resolution mapping of the *TFA* loci. Plant Cell Rep 36: 611–620
8. Hisano H, Sato K (2016) Genomic regions responsible for amenability to *Agrobacterium*-mediated transformation in barley. Sci Rep 6: 37505
9. Gerasimova SV, Korotkova AM, Hertig C, Hiekel S, Hoffie R, Budhagatapalli N, Otto I, Hensel G, Shumny VK, Kochetov AV, Kumlehn J, Khlestkina EK (2018) Targeted genome modification in protoplasts of a highly regenerable Siberian barley cultivar using RNA-guided Cas9 endonuclease. Vavilov J Genet Breed 22:1033–1039
10. Hiranniramol K, Chen Y, Liu W, Wang X (2020) Generalizable sgRNA design for improved CRISPR/Cas9 editing efficiency. Bioinformatics 36:2684–2689
11. Kumlehn J, Pietralla J, Hensel G, Pacher M, Puchta H (2018) The CRISPR/Cas revolution continues: from efficient gene editing for crop breeding to plant synthetic biology. J Int Plant Biol 60:1127–1154
12. Kumlehn J, Serazetdinova L, Hensel G, Becker D, Lörz H (2006) Genetic transformation of barley (*Hordeum vulgare* L.) via infection of androgenetic pollen cultures with *Agrobacterium tumefaciens*. Plant Biotechnol J 4:251–261

Chapter 15

In Planta Genome Editing in Commercial Wheat Varieties: Use of *TaQsd1* to Lengthen Seed Dormancy

Weifeng Luo, Yuelin Liu, and Ryozo Imai

Abstract

Dependency on in vitro culture and regeneration limits the ability to use genome editing on elite wheat (*Triticum aestivum* L.) varieties. We recently developed an *in planta* particle bombardment (iPB) technique for gene editing in wheat that utilizes shoot apical meristems (SAMs) as a target tissue. Since the method does not require in vitro culture, it can therefore be used on recalcitrant varieties. In this chapter, we describe in detail the steps used in the iPB method. With this protocol, 3% to 5% of T0 plants grown from bombarded SAMs typically carry mutant alleles and approximately 1% to 2% of the T0 plants inherit mutant alleles in the next generation.

Key words Bombardment, Genome editing, Shoot apical meristem, Wheat

1 Introduction

The development of genome editing using clustered regularly interspaced short palindromic repeats (CRISPR) and CRISPR-associated protein9 (Cas9) nuclease has made it possible to modify traits of interest in target crops in a precise manner, which has greatly accelerated modern crop breeding efforts [1–3]. Biolistic bombardment and Agrobacterium-mediated transfer are two popular methods used to introduce a CRISPR/Cas9 expression vector into plant cells. These conventional culture-based transformation methods, however, can only be used in genotypes that are amenable to cell culture and regeneration. This significant limitation prevents the application of genome editing to many commercial varieties of major crops, including important varieties of wheat (*Triticum aestivum* L.), maize (*Zea mays* L.), and soybean (*Glycine max* L.). Therefore, an *in planta* particle bombardment (iPB) method that utilizes the shoot apical meristem (SAM) as a target tissue for genome editing was developed to circumvent these limitations [4, 5].

Naoto Kawakami and Kazuhiro Sato (eds.), *Seed Dormancy: Methods and Protocols*, Methods in Molecular Biology, vol. 2830, https://doi.org/10.1007/978-1-0716-3965-8_15,

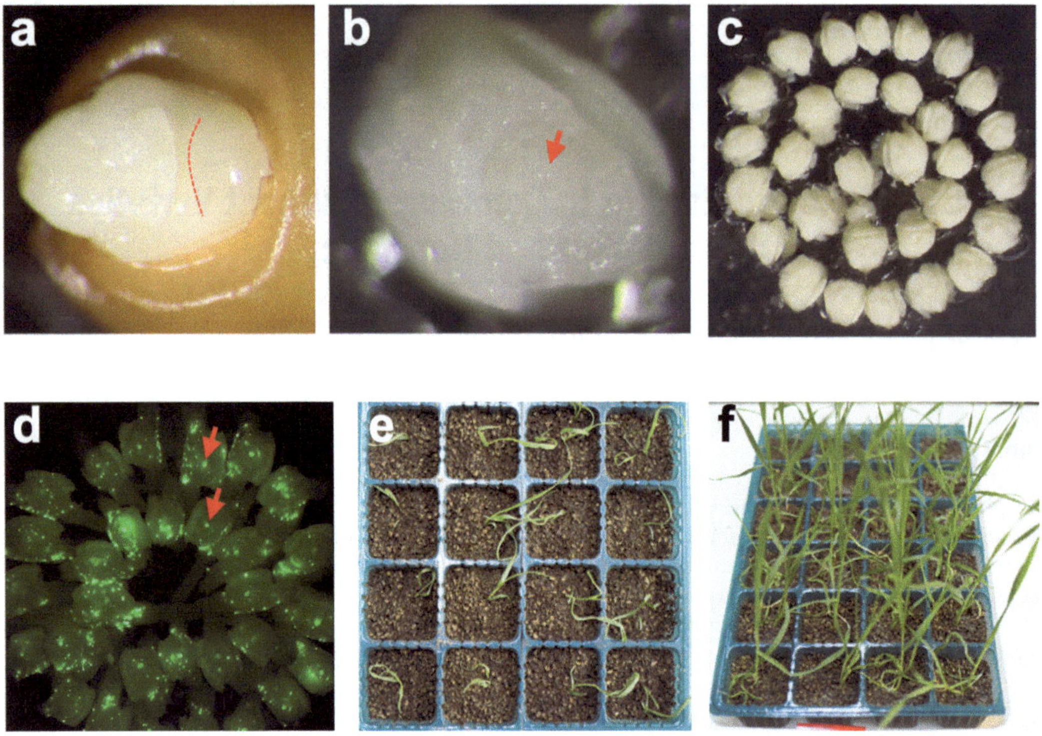

Fig. 1 The iPB procedure used for genome editing in wheat. (**a**) Coleoptiles and the first three leaves are removed from mature embryos under a microscope. (**b**) Image of an excised embryo under bright field; red arrow points to the SAM; (**c**) SAMs are arranged on a MS plate prior to bombardment; (**d**) Fluorescence image of the whole apical tissues 12 h after bombardment; the red arrows point to SAMs; (**e**) Bombarded wheat plants 2–3 weeks after being transferred to soil. (**f**) Wheat plants are grown under long-day conditions until the fifth leaf develops

SAMs are composed of undifferentiated cells that eventually develop into leaf and floral organs. SAMs are organized in three cells layers: L1, L2, and L3. The L1 and L3 layers are destined to develop into epidermal and vascular cells, respectively. Germ cells such as pollen and embryo sacs are derived from L2 cells in both dicots and monocots [6, 7], which enables these cells to be used as a target for transformation and genome editing to generate inheritable mutations. In imbibed wheat seeds, the SAM of an embryo can be exposed by removing the coleoptile and the first three leaves. Foreign DNA can then be introduced to L2 cells of the SAM by particle bombardment. Stable transgenic wheat lines have been created by bombarding a GFP-expression plasmid into SAMs [8].

We use dry commercial seeds as start material. After SAM was exposed, we bombarded gold particles coated with CRISPR/Cas9 expression plasmid into SAMs (Fig. 1). We were able to screen the positive T0 plants after 4 weeks (spring cultivars) and 7 weeks (winter cultivars) (Fig. 2). All the seeds of positive T0 plants were collected and grown. We analyzed the first leaf of T1 plants to

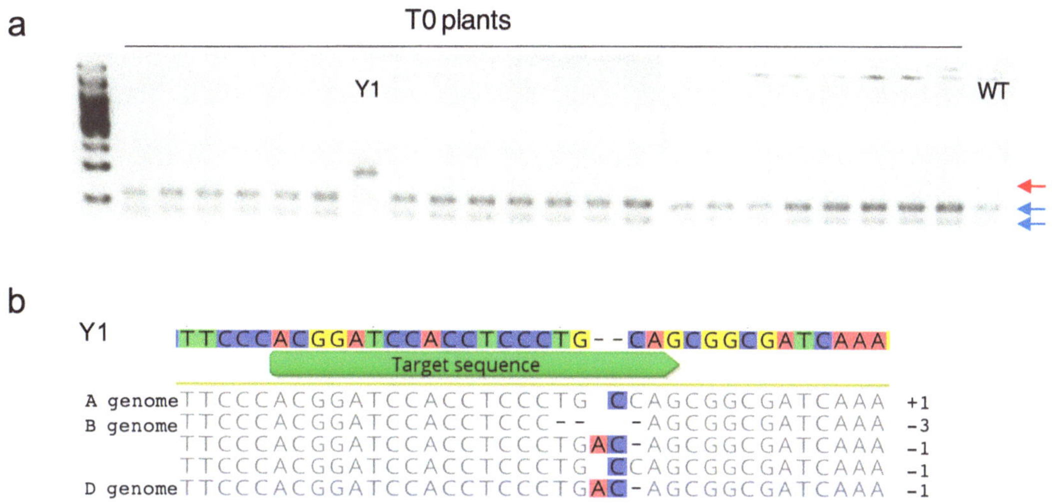

Fig. 2 Cleaved amplified polymorphic sequence (CAPS) analysis of the TaQsd1 locus in T0 plants. (**a**) Genomic DNA isolated from the fifth leaf of T0 plants was subjected to a CAPS analysis. PCR products were amplified using A, B, and D genome common primer sets. Red and blue arrows indicate undigested and digested bands after restriction enzyme digestion, respectively. A 100 bp ladder was used as a size marker. (**b**) The mutant alleles of Y1 as identified by sequencing

confirm whether the mutations were inherited to T1 generation or not (Fig. 3). With this method, we have successfully achieved genome editing in commercial wheat varieties [4, 5, 9].

In this chapter, a detailed protocol for iPB-based genome editing is described. The protocol has been utilized to create mutants of the *TaQsd1* gene, a negative regulator of seed dormancy, in a Japanese elite winter cultivar, "Yumechikara" [10].

2 Materials

2.1 Plant Materials

While the description of this protocol utilizes the winter wheat cultivar, "Yumechikara," other winter or spring wheat cultivars can also be used (*see* **Note 1**). The use of mature seeds harvested from greenhouse grown plants is recommended. Seeds harvested in the field can also be used as long as they are clean and free from fungal and bacterial contamination.

2.2 Medium and Vectors

1. Murashige and Skoog (MS) medium [11]: For 1 L medium, use 4.3 g Murashige and Skoog Basal Slat Mixture (MS) (Sigma–Aldrich, USA). Add 30 g Maltose Monohydrate (Nacalai Tesque, Japan) and 0.98 g 2-morpholinoethanesulfonic acid (MES) monohydrate (DOJINDO LABORATORIES, UAA). Adjust the pH to 5.8 with 1 M Potassium Hydroxide (KOH) (Wako, Japan). Add 7.0 g Phytagel (Sigma–Aldrich, USA) agar and 3 mL Plant

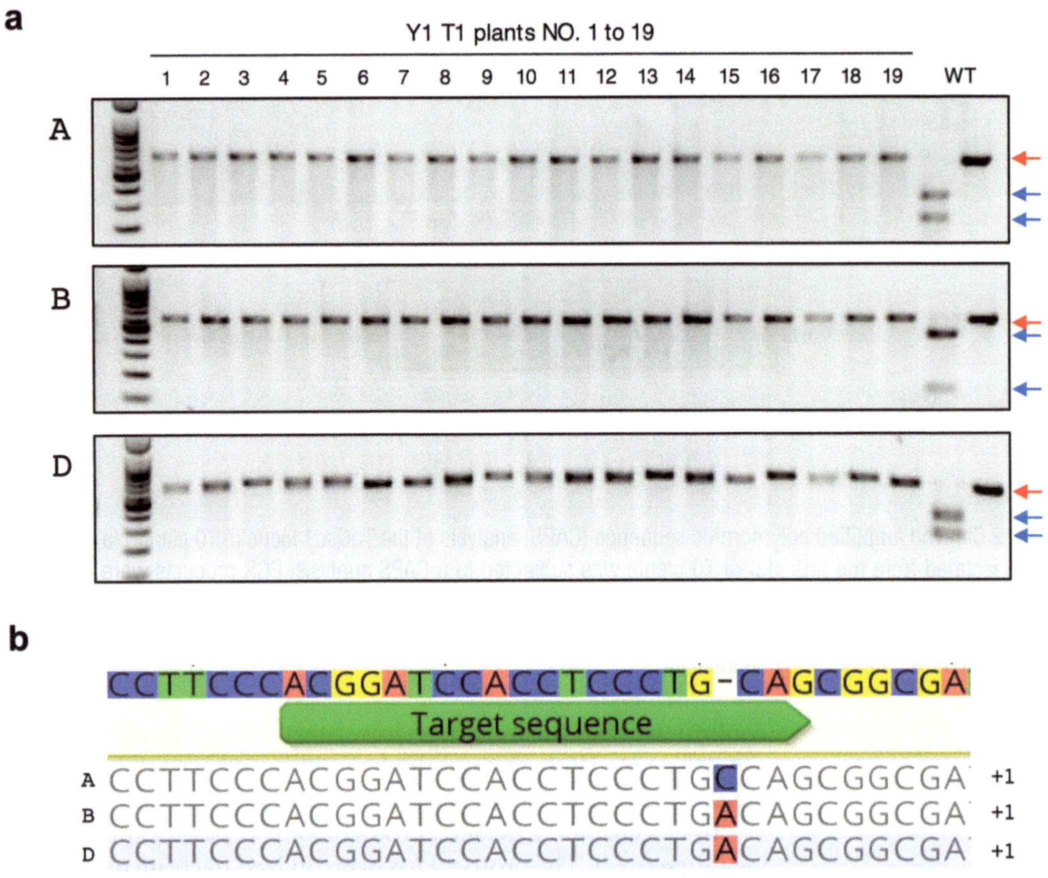

Fig. 3 Cleaved amplified polymorphic sequence (CAPS) analysis of the TaQsd1 locus in T1 plants. (**a**) Genomic DNA was isolated from the fifth leaf of Y1 plants and then subjected to a CAPS analysis. PCR products were amplified using A, B, and D genome-specific primer sets. Red and blue arrows indicate undigested and digested bands after restriction enzyme digestion, respectively. A 100 bp ladder was used as a size marker. (**b**) The genotype of Y1-15 as identified by sequencing

Preservative Mixture (PPM) (Plant Cell Technology, USA) directly into the mixture. Autoclave the prepared mixture at 121 °C for 20 min. After sterilization, cool the mixture by placing the container on a clean bench. Add 1 mL Murashige and Skoog Vitamin Solution (Sigma-Aldrich, USA) and mix the liquid medium before dispersing it into thick petri dishes (90 mm × 20 mm). The MS medium plates can then be subsequently stored at 4 °C.

2. Prepare three plasmid vectors expressing Cas9 [pE(R4-R3) ZmUbi_OsCas9_ver3], GFP (pUba-GFP), and guide RNA for *TaQsd1* (pTaU6gRNA) as previously described [4, 12] (*see* **Note 2**). The use of pUba-GFP is optional (*see* **Note 3**).

3. Plasmid DNA (1 mg/mL): To obtain high-quality DNA, prepare plasmids using a RealGenome plasmid purification kit (RBC Bioscience, Taiwan) or a similar kit (*see* **Note 4**).

2.3 Components for Particle Bombardment

1. The protocol utilizes 0.6 μm gold particles (Bio-Rad Laboratories Ltd., UK), which work well for SAM bombardment. Macrocarriers, macrocarrier holders, stopping screens, 1350 psi, or 650 psi (pounds per square inch) rupture discs can all be purchased from Bio-Rad (*see* **Note 5**) (all Bio-Rad Laboratories Ltd., UK).
2. 0.1 M Spermidine (free base): First prepare a 1 M solution by dissolving 1 g spermidine powder in 6.89 mL sterile distilled water. Filter-sterilize (0.22-μm filter) the prepared solution on a clean bench. Aliquot the prepared spermidine solution into 50 μL volumes and immediately store at −80 °C. For the working stock, dilute a 50 μL aliquot of 1 M spermidine with 450 μL sterile distilled water to make 0.1 M solution. Aliquot the diluted solution into 100 μL volumes and store immediately at −30 °C.
3. 2.5 M Calcium chloride (Fisher Scientific UK Ltd., UK): Dissolve 3.67 g $CaCl_2{\cdot}2H_2O$ in 10 mL water. Mix well by vortexing. Filter-sterilize (0.22-μm filter) the prepared solution on a clean bench. Aliquot the prepared solution into 100 μL volumes and store at −30 °C.

2.4 Equipment and Related Supplies

1. Helium particle gun: Biolistic® PDS-1000/He particle delivery system (Bio-Rad).
2. Microscope (SZX16, OLYMPUS, Japan).
3. Needle (ø0.2 mm; Type A, 34G, 4 mm; TERUMO, Japan).
4. Scalpel (#12; KAI medical, Japan).
5. Micropore™ Surgical Tape (1530-0; 3M, USA).
6. Tweezer (K-7; KFI, Japan).
7. Low-binding type Microfuge tube (BMBio, Japan).
8. Microtube mixer (MT-360; TOMY, Japan).
9. Ethanol analytical grade.

3 Methods

3.1 Exposure and Preparation of SAMs

1. Sterilize dry wheat seeds by soaking in commercial bleach (6.0% sodium hypochlorite) for 20 min.
2. Rinse three times in sterile distilled water.
3. Transfer seeds to a petri dish containing a filter paper moistened with sterile, distilled water (*see* **Note 6**).

4. Geminate seeds at 4 °C for 48 h (*see* **Note 7**).
5. Remove seed coats under a stereomicroscope with the aid of a tweezer.
6. Remove the parts of the coleoptile with a needle under a stereomicroscope following the red line in Fig. 1a (*see* **Note 8**).
7. Remove first three leaves covering the SAM under a stereomicroscope with the aid of a needle (*see* **Note 9**).
8. Cut the embryo off from the endosperm under stereomicroscope using a scalpel (Fig. 1b).
9. Subsequently place SAMs upright on a MS medium plate. Each petri plate should contain approximately 30 embryos per plate arranged in a circle (diameter 1.0 cm) (Fig. 1c).

3.2 Preparation of Particle Bombardment Solution

1. Weigh 45 mg gold particle (0.6 μm) into a low-binding microfuge tube.
2. Add 500 μL 70% EtOH.
3. Vortex 5 min (set mixing speed between 6 and 10).
4. Sit for 15 min at room temperature.
5. Pellet gold by centrifugation at 13,000× *g* for 5 s.
6. Remove supernatant, and resuspend with 500 μL nuclease-free water, then vortex for 1 min with a vortex mixer.
7. Sit for 1 min at room temperature, and centrifuge at 13,000× g for 5 s.
8. Repeat **steps 6** and 7 three times.
9. Discard supernatant and add 500 μL RNase-free water and store at 4 °C (for up to 1 month).

3.3 Preparation for Bombardment

This described mixture allows for a total of four bombardments. The experiment is performed under sterile conditions, and the tubes are kept on ice during the whole procedure.

1. Vortex the gold particle solution thoroughly and resuspend by ultrasonication for 5 s. Then, transfer 15 μL gold particle solution to a low-binding microfuge tube.
2. Add 3 μL of the Cas9 expression plasmid, 3 μL of the gRNA expression plasmid, and 3 μL of the GFP expression vector, and mix with a pipette.
3. Add 25 μL 2.5 M $CaCl_2$ to the mixture (*see* **Note 10**).
4. Add 10 μL 0.1 M spermidine to the mixture (*see* **Note 10**).
5. Mix the prepared solution for 5 min (set mix speed between 3 and 4).
6. Place the mixture at room temperature for 10 min to allow the plasmids to precipitate onto the gold particles.

7. Pellet gold particles by centrifugation at 9000× *g* for 5 s.
8. Discard the supernatant.
9. Add 70 μL 70% ethanol and resuspend the gold particles by vortexing, followed by centrifugation of the tube at 9000× *g* for 5 s.
10. Discard the supernatant.
11. Repeat **steps 9** and **10**.
12. Add 26 μL 100% ethanol (analytical grade) and resuspend the gold particles by placing the tube in an ultrasonic sonicator bath for 5–10 s.
13. Immediately apply 6 μL of the suspension to the center of the microcarrier and spread evenly. Allow to the microcarrier to dry.

3.4 Microprojectile Bombardment

1. The Biolistic PDS-1000/He particle delivery system should be preassembled according to the manufacturer's instructions. The device should be situated in a laminar flow hood.
2. In the described procedure, we used a 6 mm gap distance between the helium nozzle and the launch assembly, and a 6 cm distance from the stopping screen to the SAMs. The vacuum in the chamber was 27 inches of Hg. We stacked two 650 psi rupture discs for a single bombardment (*see* **Note 11**). Each plate is bombarded a total of 4 times.

3.5 Postbombardment Incubation

1. Successful delivery of plasmids into the SAM can be observed by GFP fluorescence 12 h after a bombardment (Fig. 1d) (*see* **Note 12**).
2. Transfer SAMs into a new petri plate containing MS medium and incubate for 2–3 weeks under long-day conditions (16 h light/8 h darkness) at 22 °C and then transfer the developing wheat plants to soil (Fig. 1e).
3. After 2 weeks, transfer all wheat plants to a 4 °C incubator and grow for 5 weeks under short-day conditions to complete the vernalization requirement (*see* **Note 13**).
4. Transfer all wheat plants to a 22 °C incubator and incubate under long-day conditions (Fig. 1f).

3.6 Mutant Screening

1. Extract genomic DNA from fifth leaf and perform a cleaved amplified polymorphic sequence (CAPS) analysis to screen T0 mutant plants (Fig. 2).
2. Harvest the seeds from selected T0 plants and grow T1 plants.
3. Extract the genomic DNA from first leaf of the T1 plants and confirm genotypes by a CAPS analysis (Fig. 3).

4 Notes

1. Wheat seeds are usually germinated in a petri dish containing moistened filter paper either at 4 °C for 48 h or at 22 °C overnight. Incubation of winter wheat seeds requires a longer time period, which is acceptable.
2. Other genes can also be edited using specifically designed gRNA expression vectors.
3. Use of the GFP vector (pUba-GFP) is optional. GFP fluorescence serves as a positive signal for successful bombardment. The inclusion of this vector can be omitted once the protocol has been effectively established.
4. The use of good quality and purity DNA provides a high coating efficiency. A final plasmid concentration of 1 mg/mL is recommended.
5. Optimal-sized gold particles for wheat cultivars should be utilized. Rupture discs are purchased from Bio-Rad Laboratory Ltd. We utilize 1350 psi rupture disc for SAM bombardment based on our previous data. We also combine two 650 psi rupture discs for SAM bombardment when 1350 psi rupture discs are not available.
6. The use of sterilized filter papers is recommended.
7. We used the "Yumechikara" wheat cultivar in the described protocol, so seeds were germinated at 4 °C for 48 h.
8. Removal of the coleoptile using the needle requires great care to ensure that the SAM is not damaged or completely removed.
9. Care must also be used when removing the first three leaves with the aid of the needle. Again, the SAM may be injured, in which case the SAM may die. The survival rate of the SAM following bombardment will also decrease.
10. It is important to add $CaCl_2$ and spermidine while gold particles are still suspended. This must be done quickly or the bombardment efficiency will decrease.
11. We combine two 650 psi rupture discs for bombardment.
12. This step can be skipped, unless the GFP expression vector is used.
13. We recommend subjecting plants of winter wheat cultivars to vernalization conditions when plants have two to three leaves. The vernalization step can be omitted when using spring wheat cultivars.

References

1. Miao J et al (2013) Targeted mutagenesis in rice using CRISPR-Cas system. Cell Res 23: 1233–1236
2. Shan Q et al (2013) Targeted genome modification of crop plants using a CRISPR-Cas system. Nat Biotechnol 31:686–688
3. Svitashev S et al (2015) Targeted mutagenesis, precise gene editing, and site-specific gene insertion in maize using Cas9 and guide RNA. Plant Physiol 169:931–945
4. Hamada H et al (2018) Biolistic-delivery-based transient CRISPR/Cas9 expression enables in planta genome editing in wheat. Sci Rep 8:14422
5. Imai R et al (2020) *In planta* particle bombardment (IPB): a new method for plant transformation and genome editing. Plant Biotechnol 37:171–176
6. Stewart RN (1978) Ontogeny of the primary body in chimeral forms of higher plants. In: The clonal basis of development. Academic, pp 131–160
7. Satina S, Blakeslee AF, Avery AG (1940) Demonstration of the three germ layers in the shoot apex of Datura by means of induced polyploidy in periclinal chimeras. Am J Bot 27:895
8. Hamada H et al (2017) An *in planta* biolistic method for stable wheat transformation. Sci Rep 7:17573
9. Liu Y et al (2021) *In planta* genome editing in commercial wheat varieties. Front Plant Sci 12: 388
10. Abe F et al (2019) Genome-edited triple-recessive mutation alters seed dormancy in wheat. Cell Rep 28:1362–1369.e4
11. Murashige T, Skoog F (1962) A revised medium for rapid growth and bio assays with tobacco tissue cultures. Physiol Plant 15:473–497
12. Kamiya Y et al (2020) A rapid method for detection of mutations induced by CRISPR/Cas9-based genome editing in common wheat. Plant Biotechnol 37:247–251

Part V

Breeding, Practical Applications

Chapter 16

Evaluation Methods for Preharvest Sprouting Resistance in Japanese Wheat Breeding Programs

Masaya Fujita, Nana Ashikaga, Hitoshi Matsunaka, Makiko Chono, and Shingo Nakamura

Abstract

Different methodologies have been applied for the selection of preharvest sprouting resistance in cereal breeding programs. We describe here a series of methods used in practical wheat breeding programs in Japan, including phenotyping based on germination score after artificial rain treatments and genotyping using DNA markers. These methods can be modified and applied to breeding programs in which preharvest sprouting is a problem during cereal cultivation.

Key words Wheat breeding, Seed dormancy, Preharvest sprouting, Germination test, Selection method

1 Introduction

Japan is located in the eastern part of monsoon Asia, which receives the most rainfall in the world. Indeed, Japan has an average annual precipitation of 1718 mm, or about twice the world average (880 mm). The rainy season starts in June, which coincides with the wheat (*Triticum aestivum*) harvest season, thus raising concerns over preharvest sprouting (PHS) for wheat cultivation. Therefore, wheat cultivars grown in Japan must display good resistance to PHS, which has driven the development of evaluation methods for PHS resistance in Japan to help select lines with higher PHS resistance in breeding programs.

PHS resistance levels are strongly influenced by environmental factors. For example, lower temperatures during seed development lead to greater seed dormancy, making cultivars more resistant to PHS. On the contrary, a longer after-ripening period and lower temperatures during seed hydration reduce seed dormancy and thus increase the susceptibility of cultivars to PHS. The effects from these factors are therefore critical when considering how to

Naoto Kawakami and Kazuhiro Sato (eds.), *Seed Dormancy: Methods and Protocols*, Methods in Molecular Biology, vol. 2830,
https://doi.org/10.1007/978-1-0716-3965-8_16,

assess PHS resistance of cultivars and have guided efforts to find suitable conditions for the evaluation of PHS resistance.

To evaluate PHS resistance, two methods, artificial rain treatment using intact spikes and germination tests using threshed grains, have been mainly used in breeding programs [1]. Artificial rain treatment can evaluate seed dormancy levels together with other PHS resistance factors such as morphology of spikes. Using these evaluation methods and adjusting assay conditions such as temperature and after-ripening period, at the Kitami Agriculture Experimental Station of Hokkaido Research Organization (KAES) in Japan, an excellent PHS-tolerant winter wheat line, "Kitakei 1802," was successfully developed [2]. This line hardly sprouted through 10 days' continuous rainfall treatment at low temperature (15 °C) and maintained high falling number (>300 s) in two-year trials.

In addition to PHS, there is a phenomenon that unacceptable high levels of α-amylase accumulated in ungerminated grains before harvest [3]. It is not germination, but the highly accumulated α-amylase can damage grain quality as α-amylase induced by PHS does. This phenomenon is called late-maturity a-amylase (LMA). LMA is induced by a cool temperature during grain development. Susceptivity of LMA is genetically determined, and recently, a causative gene for a major LMA quantitative trait locus (QTL) LMA-1 has been found as a putative *ent*-Copalyl diphosphate synthase [4]. To evaluate LMA resistance trait, methods measuring grain α-amylase activity have been used in breeding programs.

In addition to these phenotype-based methods, recent advances fueled by the identification of causal genes controlling seed dormancy in wheat and barley (*Hordeum vulgare*) now offer a genotype-based approach to improving PHS resistance. Specific DNA markers make it possible to determine whether a cultivar or breeding line harbors the dormant or nondormant allele at the identified causal gene. For example, *Mother of FT and TFL 1* (*MFT*) was identified as the causal gene for a wheat major seed dormancy quantitative trait locus (QTL) [5], with a single-nucleotide variation (SNV) changing a T to a C 222 bp upstream of the promoter sequence of *MFT*. Incorporating this polymorphism into other cultivars can improve their PHS resistance. This polymorphism can be easily tracked in segregating populations using DNA markers, as recently demonstrated with the breeding of the winter wheat cultivar with high preharvest sprouting resistance "Yumechikara 2020" [6] in the Hokkaido area. Recent QTL analysis has revealed that the excellent PHS resistance character of "Kitakei 1802" mainly resulted from a high level of seed dormancy caused by the (*MFT*) dormant allele and a newly found major QTL, *Qphs-2D*, on chromosome 2D [7]. Interestingly, Qphs-2D could be detected by only artificial rain treatment, and germination test

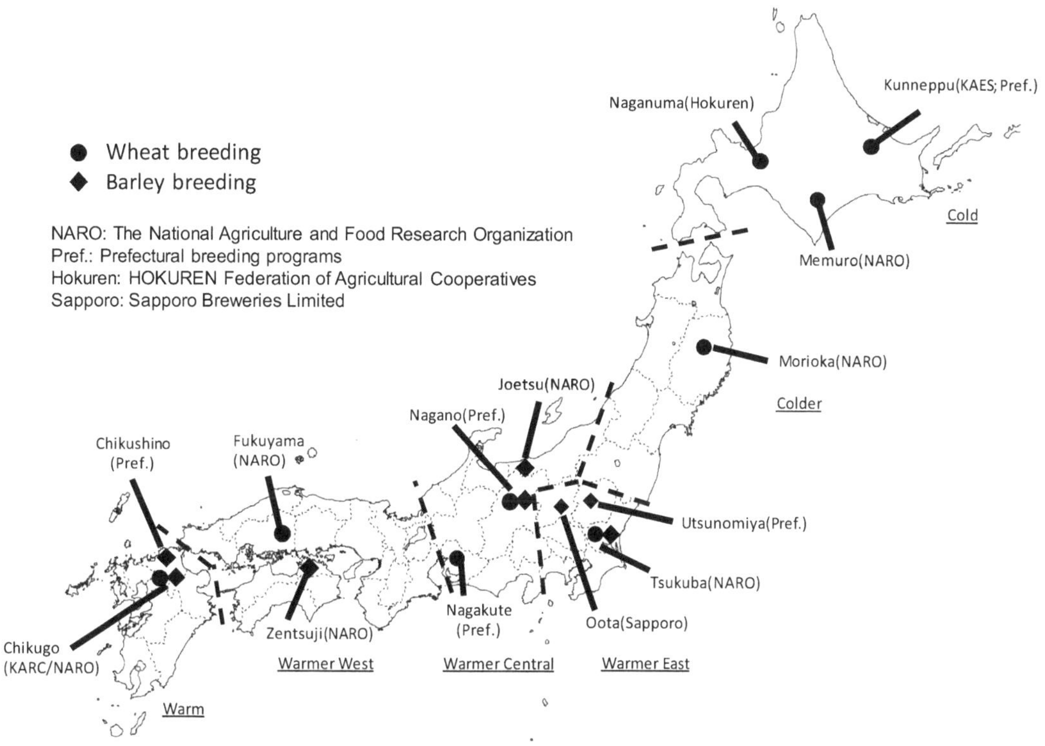

Fig. 1 Locations of the Japanese cereal breeding programs

could not detect it. This result has suggested importance of artificial rain treatment for evaluation of PHS resistance.

Other useful natural mutations have been identified from the physiological study of wheat cultivars. For instance, the sesquiterpenoid plant hormone abscisic acid (ABA) contributes to seed dormancy and germination. In addition to the dormant *MFT* allele, several Japanese cultivars carry loss-of-function mutations in the ABA catabolism gene *ABA 8′-hydroxylase* (*TaABA8′OH1*), resulting in higher ABA levels in seeds and thus lower seed germination upon harvest [8]. DNA markers for polymorphisms in ABA-related genes have been developed and facilitate the evaluation of PHS resistance in wheat.

Japan has a wide range of latitudes and requires different breeding programs for the varied crop growing conditions (Fig. 1). We summarize here several PHS evaluation methods in Japanese wheat and barley breeding programs implemented in different areas and for distinct crops (Table 1) [9]. In this article, we'd like to focus on the evaluation methods used at KAES as a typical example of evaluation methods in Japan. In addition, we mention the methods used at the Kyushu Okinawa Agricultural Research Center, National Agriculture and Food Research Organization (KARC/NARO) in the Notes section.

Table 1
Summary of evaluation methods of preharvest sprouting tolerance in representative Japanese wheat (A) and barley (B) breeding programs. Methods are summarized from Fujita (2019) [9]

(A) Evaluation of wheat breeding programs							
Region	**Cold (winter)**	**Cold (spring)**	**Colder**	**Warmer East**	**Warmer Central**	**Warmer West**	**Warm**
location	Kunneppu	Kunneppu	Morioka	Tsukuba	Nagano	Fukuyama	Chikugo
Rain shelter	None	None	None	None	None	None	After heading
Harvest period	Maturity, 7 days later	maturity, 10 days later	Maturity	Maturity	Maturity	Maturity, 7 days later	Maturity
Drying condition	40 °C 24 h	40 °C 24 h	Air-dry 1, 5 days	Air-dry 7, 14, 28 days	Air-dry 24 h	35 °C 24 h	35 °C overnight
Dormancy retention	−20 °C	−20 °C	None	None	None	−20 °C	−25 °C
Sample	Spike	Spike	Spike	Spike	Spike	Spike	Spike (test by grain)
Soaking (prewetting)	None	None	24 h	None	Overnight	4 h	None
Wet duration	7–10 days	7–10 days	7 days	7 days	7 days	7 days	7 days
Wet condition	Wet chamber	Wet chamber	Mist chamber	Watering and plastic cover	Wet chamber	Wet chamber	Petri dish 6 mL water
Temperature	15 °C	15 °C	15 °C	10 °C, 15 °C	18 °C	15 °C	10 °C, 15 °C, 20 °C
Replication	10 spikes	10 spikes	10 spikes			10 spikes	30 grains (5 spikes)
Grain germination test	Occasional	Occasional	Average 5 spikes	Occasional	Advanced lines	Total, germinated grain	30 grains
Grading	0–5 score (with controls)	0–5 score (with controls)	With controls	With controls	With controls	S-VT with controls	VS-VT with controls
VS (controls)*							Minaminokaori
S		Haruyutaka	Kitakamikomugi			Minaminokaori	
MS	Chihokukomugi		Aobakomugi		Shiranekomugi	Shiroganekomugi	
M	Hokushin	Haruyokoi		Norin 61		Norin 61	Shiroganekomugi
MT	Kitahonami		Nanbukomugi		Yumekaori		
T	Kitakei 1354	Harukirari	Nebarigoshi	Minaminokomugi	Kinuhime	Zenkoujikomugi	Norin 61
VT	OW 104, Kitakei 1838	OS71, 20S24	Ginganochikara	Zenkoujikomugi	Zenkoujikomugi		Zenkojikomugi

(B) Evaluation of barley breeding programs

Region	Warmer East	Warmer Central	Warmer West	Warm
location	Utsunomiya	Nagano	Zentsuji	Chikugo
Rain shelter	After flowering	None	None	Full growth duration
Harvest period	Maturity	Maturity	Maturity	Maturity
Drying condition	35 °C 24 h	Air-dry 24 h	35 °C 48 h	35 °C 24 h
Dormancy retention	−20 °C (s) -70 °C (g)	None	−80 °C	−20 °C
Sample	Spike (s) and grain (g)	Spike	Grain	Grain
Soaking (prewetting)	24 h	Overnight	None	None
Wet duration	7 days	7 days	5, 7 days	3, 5, 7 days
Wet condition	Wet chamber (s), Petri dish (g)	Wet chamber	4.5 mL per dish	4.5 mL per dish
Temperature	17 °C	18 °C	17 °C	15 °C
Replication	10 spikes		Two dishes	Two dishes
Grain germination test	100 grain germination**	Advanced lines	100 grains	100 grains
Grading	11 scores to 7 grades	With controls	With controls	With controls
VS (controls)	Tochinoibuki		Morex	
S	Mikamo Golden	Shunrai		Amagi Nijo
MS			Misato Golden	
M	Sachiho Golden		Haruna Nijo	Haruna Nijo
MT			Ishukushirazu	Kawasaigoku
T	Sky Golden	Setsugenmochi	Sky Golden	Ichibanboshi
VT			Sayakaze	Kashimamugi

* *VS* very susceptible, *S* susceptible, *MS* moderately susceptible, *M* moderate, *MT* moderately tolerant, *T* tolerant, *VT* very tolerant

** Germination index = $[(7 \times n2 + 6 \times n3 + 5 \times n4 + 4 \times n5 + 3 \times n6 + 2 \times n7 + 1 \times n8)/7$ ($n1, n2, \ldots, n8$ are the numbers of grains germinated on the first, second, and subsequent days until the 8th day, respectively)

2 Materials

(Items in Subheadings 2.1, 2.2, and 2.3 are used in methods at KAES.)

2.1 Artificial Rain Treatment (PHS Test) on Intact Spikes

1. Spikes (*see* **Notes 1** and **2**). As control, prepare spikes from a set of standard varieties with known PHS resistance (Table 1) and other varieties for comparison.
2. Drying chamber set to 40 °C.
3. Freezer set to −20 °C.
4. An air-conditioned (15 °C) chamber with a spike stand (Fig. 2).
5. Wetting system for spike samples in the chamber (e.g., Fig. 2; with hand-watering and a plastic cover).

2.2 Standard Threshed Grain Germination Test (Seed Dormancy Test)

1. Five spikes.
2. Drying chamber set to 40 °C.
3. Freezer set to −20 °C.
4. Filter papers (No. 2, 8.5-cm diameter) and Petri dishes (9-cm diameter).
5. Chemichlon-G solution (calcium hypochlorite) to surface-sterilize seeds. Dissolve 1 g of Chemichlon-G in 1 L of water. Store at room temperature. Prepare the appropriate amount for 1-day use (each test sample requires 35 mL).
6. Dark chamber set to 15 °C.

Fig. 2 Artificial rain treatment at KAES

2.3 Analysis of Grain α-Amylase Activity

1. Sample mill (e.g., IWATANI, Japan).
2. Plastic bag.
3. α-Amylase extraction buffer: 0.5% (w/v) NaCl, 0.02% (w/v) $CaCl_2 \cdot 2H_2O$. Add 5 g NaCl and 2 g $CaCl_2 \cdot 2H_2O$ into a 10-L cylinder or beaker. Adjust volume to 10 L with deionized water and mix. Store at 5 °C.
4. Substrate solvent: 0.1 M phosphate, pH 5.0. Add 13.6 g of KH_2PO_4 into a 1-L cylinder. Adjust volume to 1 L with deionized water. Mix and dissolve completely. Adjust pH with 0.1 M Na_2HPO_4. Store at 5 °C.
5. Substrate: Amylase test tablets (Phadebas, Sweden).
6. Substrate suspension: Prepare fresh before each use. Dissolve one Phadebas® amylase test tablet in 4.5 mL of substrate solvent. Prepare the necessary amount for 1-day use in a 500-mL beaker (each test sample uses 1.5 mL).
7. Stopping reagent: 0.5 M NaOH. Add 20 g of NaOH to a 1-L cylinder. Add deionized water and mix. Adjust volume to 1 L with water. Store at room temperature.
8. Polypropylene tube (5-mL capacity).
9. Eight-strip tube (1.2-mL capacity) and 96-well rack.
10. Micropipette (200 μL, 1000 μL).
11. Eight-channel micropipettes (200–300 μL).
12. Large-bore tip (200 μL).
13. Water bath (for incubation).
14. Water bath with stirrer (for substrate suspension).
15. A 96-well microplate.
16. Spectrophotometer.

2.4 Fine-Tuned Grain Germination Assays

1. Plants grown in the field, greenhouse, or growth chamber.
2. Tags for marking heading date on spike.
3. Drying chamber set to 30 °C.
4. Envelopes and plastic bags to store seeds.
5. Vacuum sealer for the plastic bags.
6. Freezers set to −30 °C.
7. Filter papers (No. 2, 8.5-cm diameter).
8. Petri dishes (9-cm diameter).
9. Distilled water.
10. Dark growth chambers set to 15 °C and/or 20 °C.

2.5 Selection Based on the *MFT-3A* DNA Marker

1. Plant materials for extraction of genomic DNA (e.g., 4-day-old shoots, a young leaf).
2. An incubator to dry plant materials at 50 °C.
3. Primers.

   Forward primer: 5′-GTAGCGGGTGAAATCTGCAT-3′.

   Reverse primer: 5′-GGGACGTACGAGGGTGTAGA-3′.
4. Restriction enzyme: *Cla*I.
5. Equipment and materials for genotyping (e.g., PCR systems, agarose gel electrophoresis, staining gel, and detection system).
6. 0.5× Tris borate EDTA (TBE) buffer. Prepare 0.5× solution by diluting 10× TBE stock (1 L contains 108 g Tris base [2-amino-2-(hydroxymethyl)-1,3-propanediol], 55 g boric acid, and 40 mL of 0.5 M EDTA, pH 8.0, in deionized distilled water).

2.6 Selection Based on the *TaABA8′OH1* DNA Marker

1. Primers.

   Forward primer 5′-AAGCCCAACACGTTCATGCCGTTC-3′.

   Reverse primer 5′-CGT(C/T)GCC(A/T)CTATCG(C/T)GC(C/T)GTTGAT-3′.
2. Equipment and materials for genotyping (e.g., PCR systems, acrylamide gel electrophoresis, staining gel, and detection system).

3 Methods

(Steps in Subheadings 3.1, 3.2, and 3.3 are methods at KAES; see Table 2.)

3.1 Artificial Rain Treatment (PHS Test) on Intact Spikes

1. After harvesting, dry the spikes in the drying chamber at 40 °C for 24 h (*see* **Notes 3**, **4**, and **5**) and then store at −20 °C until use.
2. Place harvested spikes on a stand. Cover the stand with plastic film (Fig. 2).
3. Incubate the spikes at 15 °C for 7 to 10 days with checking sprouting state of standard varieties (*see* **Notes 6** and 7).
4. Sprinkle water on the spikes every 12 h to keep the spikes wet (*see* **Note 8** and Fig. 2).
5. After treatment, score germination state of the spikes on a scale of 0 (no sprouted kernels visible) to 5 (all kernels sprouted) (Fig. 3). The sprouting score is calculated by counting the number of sprouted kernels and dividing the score by 2. For

Table 2
Selection scheme for PHS-resistant spring wheat at KAES in 2020

Generation	Application	Method of evaluation	Subheading
F_3–F_4	Spike selection	Artificial rain treatment on intact spikes	2.1, 3.1
F_4–F_5	Nursery evaluation by grain samples	Standard threshed grain germination test	2.2, 3.2
F_6 or later	Selection in yield trial	Artificial rain treatment on intact spikes	2.1, 3.1
		Standard threshed grain germination test	2.2, 3.2
		Analysis of grain α-amylase activity	2.3, 3.3

Breeding lines in the yield trial (F_6 or later self-pollinated generations) were grown at two locations in Hokkaido (Central AES and KAES), and their PHS resistance was evaluated with artificial rain treatment. Fig. 5 shows an example of the evaluation data obtained by the artificial rain treatment

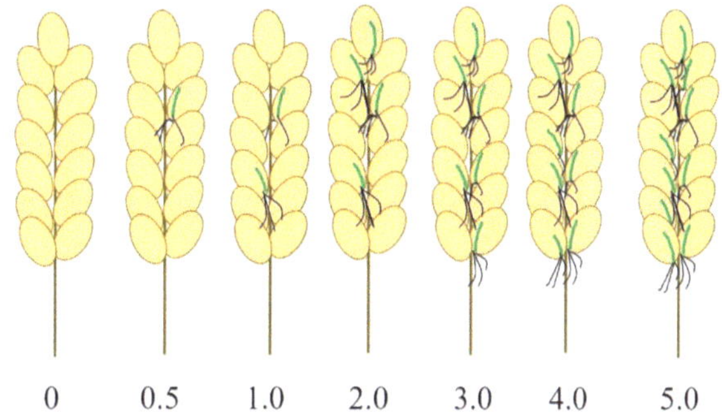

Fig. 3 Determination of sprouting score at KAES. Shoots are shown in green, and roots are shown in black. The spikes are scored on a scale from 0 (no visually sprouted kernels) to 5 (almost all kernels have sprouted). The sprouting score is calculated by counting the number of sprouted kernels and dividing it the sore by 2. If 10 or more sprouted kernels are observed, the score is set to 5

example, four sprouted kernels mean a score of 2. If 10 or more sprouted kernels are found, the score is 5.

6. Compare the sprouting score of the spikes with the scores of standard varieties to evaluate their PHS resistance (*see* **Note 9**). An example of the classification of PHS resistance is shown in Fig. 4, and the results of PHS evaluation in KAES and Central AES of Hokkaido are shown in Fig. 5.

3.2 Standard Threshed Grain Germination Test (Seed Dormancy Test)

1. Use five spikes for each sample.
2. After drying at 40 °C for 24 h, store at −20 °C until the germination test.
3. Gently thresh the individual spike to maintain the original dormancy level.

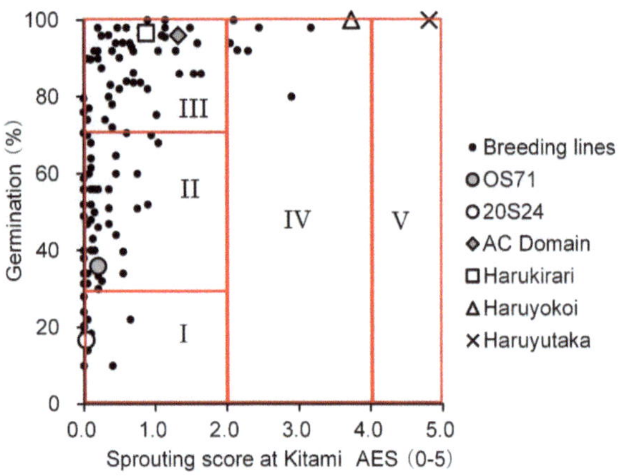

Fig. 4 Classification of PHS resistance using sprouting score and germination rate in 2019
I: Extreme resistance; II: High resistance; III: Resistance; IV: Intermediate; V: Susceptibility

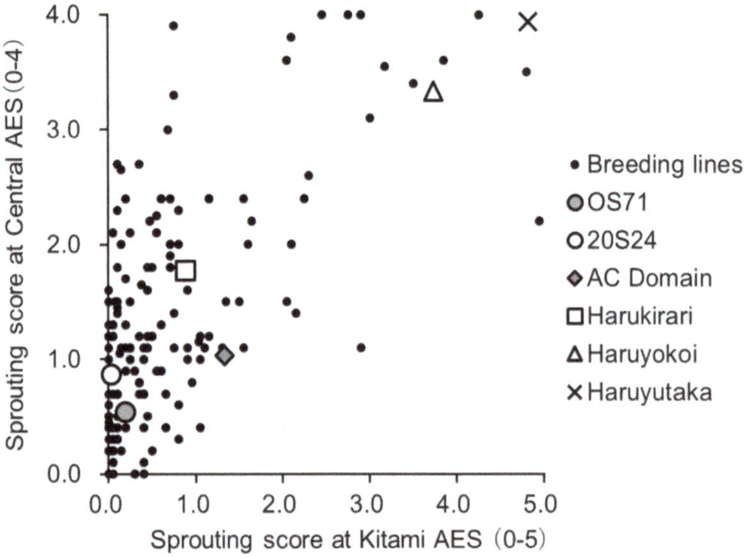

Fig. 5 Sprouting scores at the KAES and Central AES locations of Hokkaido in 2019. Central AES uses a slightly different sprouting scoring system than KAES. These samples were harvested 10 days after maturity. The artificial rain treatments were conducted for 7 days at 15 °C at KAES and for 7 to 8 days at 15 °C at Central AES

4. Use 50 grains and place them on filter paper in a Petri dish (*see* **Note 10**).
5. Surface-sterilize the grains. Place 50 grains in a tea strainer, wash them under running water for a few seconds, and drain. Return the sample to the Petri dish and add Chemichlon-G

solution until the entire sample is almost fully submerged. Incubate for 30 min. Drain the Chemichlon-G solution with a tea strainer, wash the samples and the inside of the Petri dish with running water, and drain again, then add 6 mL of water.

6. Place the Petri dishes in a dark chamber set to 15 °C (*see* **Note 11**).
7. After 3 days, count and remove germinated seeds. After 4 days, count the germinated and nongerminated seeds (*see* **Note 12**). Calculate the germination rate.
8. Classify the level of PHS resistance as shown in Fig. 4.

3.3 Analysis of Grain α-Amylase Activity (see Also Fig. 6)

1. Dry the spike samples for analysis in the drying chamber at 40 ° C for 24 h and then store at room temperature.
2. Thresh the samples individually and transfer the grains into a plastic bag or an envelope.
3. Grind 10 g of each grain sample in a mill for 1 min and then transfer the whole-wheat flour into a plastic bag (*see* **Note 13**).
4. Weigh 200 mg of whole-wheat flour into a 5-mL tube.
5. Add 4 mL of α-amylase extraction buffer to each tube and mix it for a few seconds using a vortex mixer.
6. Leave overnight at 4 °C for sedimentation of whole-wheat flour.
7. Transfer the 0.6-mL aliquot of supernatant solution into a 1.2-mL tube. Use 0.6 mL of α-amylase extract solution instead of the supernatant to prepare two blanks.
8. Preincubate the 1.2-mL tubes and substrate suspension at 50 ° C for 5 min.
9. Add 0.2 mL of substrate suspension to each tube using a large-bore tip and an eight-channel micropipette (*see* **Notes 14** and **15**).
10. Incubate the tubes at 50 °C for 30 min. Mix the tubes by flipping them upside down repeatedly every 10 min during the incubation.
11. Add 0.2 mL of stopping reagent to each tube 30 min from the time of addition of substrate suspension.
12. Centrifuge the tubes at 2000 rpm for 5 min at room temperature.
13. Transfer 0.2 mL of supernatant solution into a 96-well microplate.
14. Measure the absorbance of the supernatant at 620 nm and normalize against the blanks.
15. Calculate α-amylase activity following the instruction manual of the Phadebas® amylase test tablet:

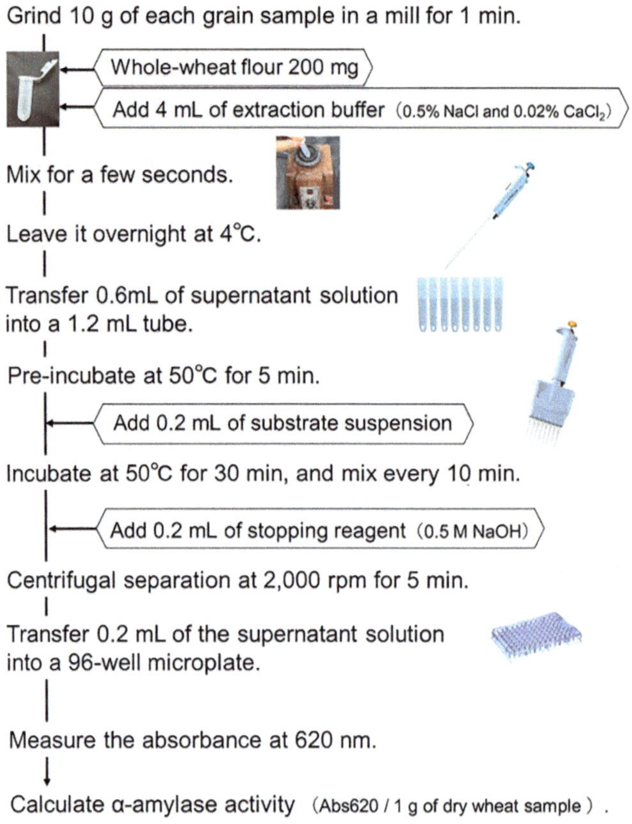

Fig. 6 Schematic workflow of α-amylase activity determination in wheat grains using blue starch as substrate. (Modified from Nakatsu [11])

Activity (μkat/L) = e^N.

$N = A + \sqrt{}\,(B + C/\ln(\text{Abs620}))$.

A, B, and C values are shown in the manual.

Abs620: Subtract the blank value from the sample value.

To express in U/L, multiply by 60. U/I = 60 × μkat/L.

16. A value of U/L = 645 or greater means that the Falling Number (FN) of the sample is under 300 s.

3.4 Fine-Tuned Grain Germination Assays

1. Grow plants in the field and check the growth conditions.
2. Mark spikes for pollination (flowering) dates with tags to collect seeds at the same maturation stage.
3. Harvest the tagged spikes in 3- to 5-day intervals from 40 to 60 days after pollination (DAP). Dry the harvested spikes at 30 °C for 2 days and collect 16–20 seeds per spike (only first and second florets) from the center of each spike. Place the dry

Fig. 7 Germination assay. Entire seeds were placed in a Petri dish (9-cm diameter) containing two Whatman No. 2 filter papers and 6 mL of water. Dishes were placed in a growth cabinet in the dark

seeds in an envelope, and place the envelope in a plastic bag. Vacuum-seal the plastic bag and store at −30 °C until use.

4. Place two No. 2 filter papers (Whatman 8.5-cm diameter) in a Petri dish (9-cm diameter) and soak them with 6 mL of distilled water. Place 30 seeds on the wet filter papers. Prepare three replicates.
5. Place the dishes in the dark at 15 °C and/or 20 °C.
6. Count germinated seeds (germination is defined as pericarp rupture over the embryo) every 24 h for 7 days and remove germinated seeds from the dishes each day (Fig. 7).
7. Calculate germination rate daily to show germinated pattern (Figs. 8 and 9).

3.5 Selection Based on the *MFT-3A* DNA Marker

1. Completely dry plant materials by incubation overnight or longer at 50 °C.
2. Extract genomic DNA from leaves using a commercial kit, such as the DNeasy Plant Mini Kit, or other method such as that

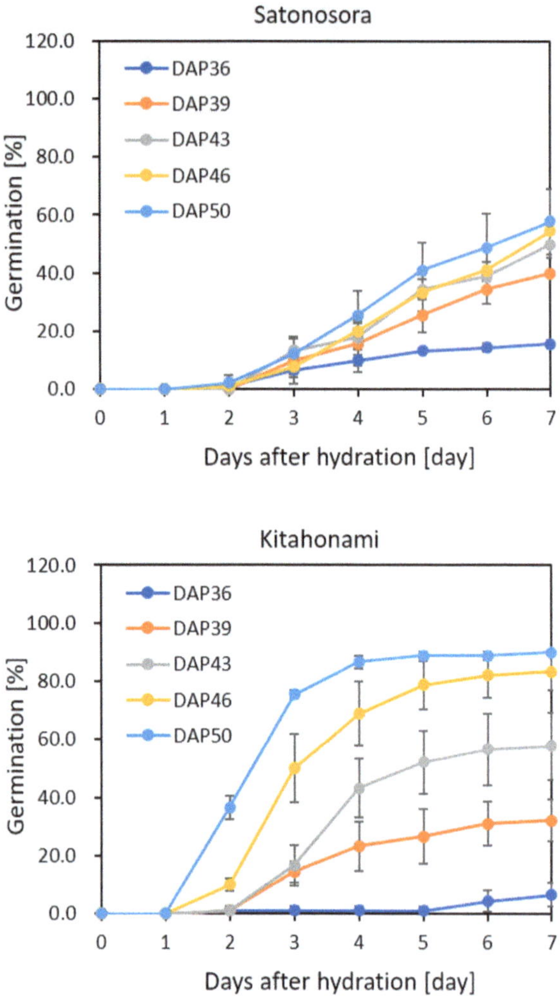

Fig. 8 Germination patterns of "Satonosora" (top) and "Kitahonami" (bottom) seeds at different seed maturation stages. Satonosora and Kitahonami seeds were harvested at 36 to 50 DAP and were hydrated at 15 °C for 7 days. Compared to Kitahonami, which carries the nondormant *MFT* allele, Satonosora showed a strong seed dormancy after physiological maturity

described in Chap. 6. Store the extracted DNA at −30 °C until use.

3. Amplify DNA fragments by PCR with specific primers using a commercial DNA polymerase according to the manufacturer's protocol. The exact PCR conditions will depend on the DNA polymerase, but an exemplar PCR protocol should consist of 40 cycles at 98 °C for 10 s, 60 °C for 15 s, and 68 °C for 1 min.
4. The amplified fragments (~800 bp in length) are digested with the *Cla*I restriction enzyme (*see* **Note 16**).

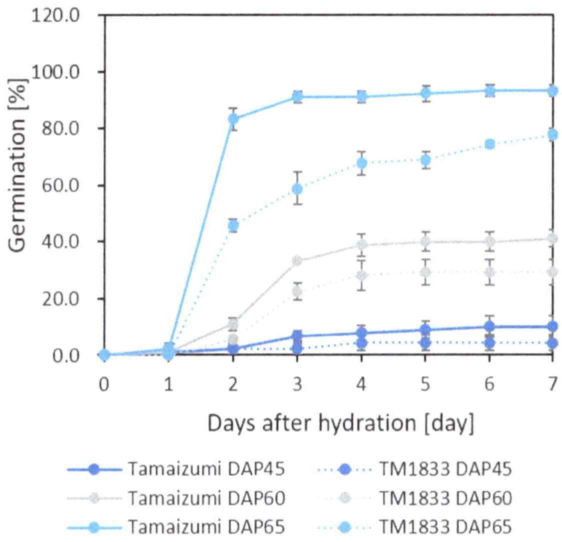

Fig. 9 Germination patterns of "Tamaizumi" and "TM1833" seeds at different seed maturation stages. TM1833 was identified from a gamma-ray-irradiated Tamaizumi line. TM1833 (a double mutant lacking TaABA8′OH1-A and TaABA8′ OH1-D function) showed higher ABA contents in embryos during seed development and lower germination than Tamaizumi (which is a single mutant for *TaABA8′OH1-D*). Tamaizumi and TM1833 seeds harvested at 45, 60, and 65 DAP were hydrated at 20 °C for 7 days

5. The restriction digests are analyzed by 1.5% (w/v) agarose gel electrophoresis in 0.5× TBE buffer.
6. Amplified fragments with the non-Zenkojikomugi-type polymorphism associated with weak dormancy harbor the *Cla*I restriction site and thus produce two fragments of approximately 400 bp each when digested. Amplicons with the Zenkojikomugi-type polymorphism for strong dormancy cannot be cut by *Cla*I due to the nucleotide substitution at −222 bp in the *MFT-3A* promoter (Fig. 10).

3.6 Selection Based on the TaABA8′OH1 DNA Marker [8]

1. Extract genomic DNA as above (Subheading 3.5).
2. Amplify DNA fragments by PCR with specific primers using a commercial DNA polymerase according to the manufacturer's protocol (e.g., 30 s denaturation at 98 °C, followed by 40 cycles of 10 s at 98 °C, 30 s at 60 °C, and 30 s at 72 °C).
3. Separate the resulting amplicons by electrophoresis and check size (e.g., amplicons from the "Norin 61" cultivar will be 246 bp (A genome), 280 bp (B genome), and 295 bp (D genome)) (Fig. 11).

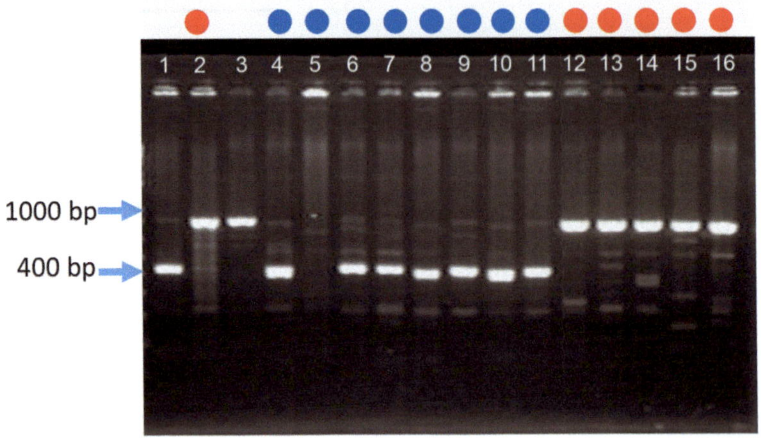

Fig. 10 Genotyping of *MFT* on chromosome 3A in various wheat cultivars using a specific cleaved amplified polymorphic sequence (CAPS) marker. Blue circles: cultivars in the northern part of Japan, Hokkaido. Red circles: cultivars in the southern part of Japan. Cultivars or lines: 1, Chinese Spring; 2, Zenkoujikomugi; 3, OS21–5; 4, Taisetsukomugi; 5, Chihokukomugi; 6, KK1354; 7, Kitamoe; 8, Hokushin; 9, Kitahonami; 10, Haruhikari; 11, Haruyutaka; 12, Norin61; 13, Shiroganekomugi; 14, Igachikugo-Oregon; 15, Minaminokomugi; 16, Shiranekomugi. Chihokukomugi shows an absence of bands due to a transposon insertion (*see* **Note 16**). Amplicons were resolved on a 1.5% (w/v) agarose gel in 0.5× TBE buffer

4 Notes

1. These varieties should be grown in the same field at the same time.
2. In the warm region, where KARC/NARO is located, the ripening period occurs almost at the same time as the beginning of the rainy season; therefore, to avoid changes in seed dormancy caused by rainfall, the seeds are sheltered from rain starting from heading day onward (Fig. 12). We attempted to maintain a constant temperature in the ripening environment. In addition, the plants are irrigated as needed to avoid drought stress.
3. At KARC/NARO, dry the spikes at room temperature for about 2 weeks.
4. After drying, check the condition of the samples. If still wet, drying should continue.
5. If plants are dried under the roof outside too long, their seeds may lose dormancy and PHS resistance may not be reliably evaluated. At KARC/NARO, samples should be evaluated within 2 weeks of drying. Samples are evaluated within 1 month, even if they are stored frozen after drying.

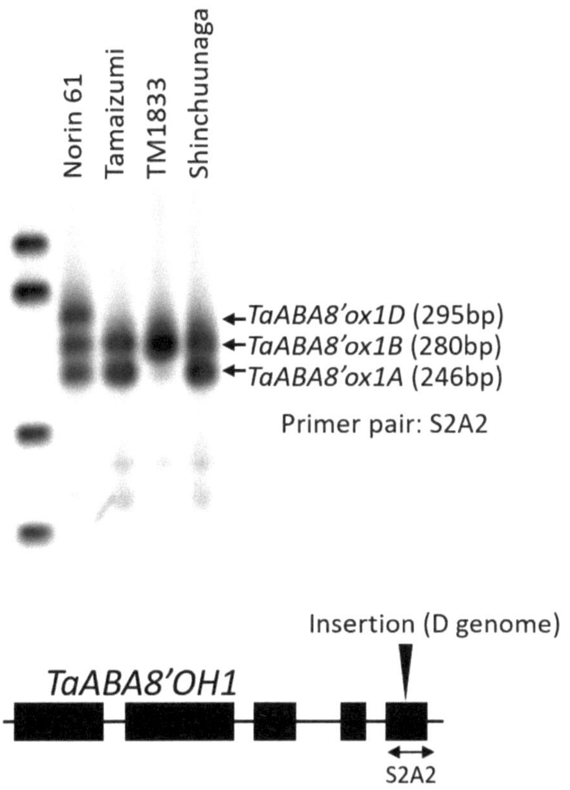

Fig. 11 Genotyping of *TaABA8′OH1* using a gene-specific primer pair. PCR was performed with a pair of gene-specific primers (S2 primer sequence 5′-AAGCCCAACACGTTCATGCCGTTC-3′, A2 primer sequence 5′-CGT(C/T)GCC(A/T)CTATCG(C/T)GC(C/T)GTTGAT-3′), designed to amplify around the stop codon, with Ex Taq DNA polymerase (Takara Bio Inc.), following the manufacturer's instructions. The PCR conditions were as follows: 30 s denaturation at 98 °C, followed by 40 cycles of 10 s at 98 °C, 30 s at 60 °C, and 30 s at 72 °C. The amplified fragments were subjected to polyacrylamide gel electrophoresis without denaturing treatment. The fragment sizes derived from the A, B, and D genomes of Norin 61 are 246, 280, and 295 bp, respectively. The fragment derived from the D genome was not amplified in Tamaizumi or Shinchuunga, as they carry an insertion in *TaABA8′OH1-D*. The fragments derived from the A and D genomes were not amplified in TM1833, which harbors a deletion in *TaABA8′OH1-A* and an insertion in *TaABA8′OH1-D*

6. An additional artificial rain treatment of 24 h at 10 °C is performed before the 15 °C treatment for samples that may have deep seed dormancy or for crosses that are expected to have high PHS resistance.
7. Check sprouting of the standard varieties and then decide the date of the assessment.
8. A showerhead is used to water.

Fig. 12 The growing area is covered with a plastic greenhouse to prevent the effects of rainfall at KARC/NARO

9. The spikes that are clearly sprouting more than the target resistant varieties can be rejected without counting the score.
10. Choose 50 healthy kernels excluding Fusarium head blight (FHB)-damaged, immature, or germinated kernels.
11. At KARC/NARO, the test temperatures for germination tests are 20 °C and 15 °C. Extremely resistant strains usually show almost no germination at 15 °C, depending on the planting season. For this reason, we conduct germination tests at 10 °C for lines that do not germinate at 15 °C.
12. Remove any FHB-damaged kernels.
13. Use two or three mills alternately to prevent overheating. Clean the mills with an air compressor.
14. Always stir the substrate suspension using a water bath with a stirrer.
15. Use large-bore tips for substrate suspension to avoid clogging tips. Dispense the substrate suspension in regular time intervals to achieve a reaction time of 30 min for all samples.
16. Detailed sequence variation for *MFT-3A* from 694 cultivars, consisting of 117 Japanese cultivars and 557 cultivars from a wheat core collection, has been described by Nakamura et al. [10]. Due to a transposon insertion in the promoter, PCR amplification is not possible in the wheat cultivar "Chihokukomugi," which has weak seed dormancy and experienced devastating damage (causing approximately 40% decrease in yield) by PHS in 1995 in Hokkaido.

References

1. Paterson AH, Sorrells ME, Obendorf RF (1989) Methods of evaluation for preharvest sprouting resistance in wheat breeding programs. Can J Plant Sci 69:681–689
2. Yanagisawa A, Nishimura T, Amano Y et al (2005) Development of winter wheat with excellent resistance to pre-harvest sprouting and rain damage. Euphytica 143:313–318
3. Mares DJ, Mrva K (2008) Late-maturity α-amylase: Low falling number in wheat in the absence of preharvest sprouting. J Cereal Sci 47:6–17
4. Derkx A, Baumann U, Cheong J et al (2021) A major locus on wheat chromosome 7B association with late-maturity α-amylase encodes a putative *ent*-copalyl diphosphate synthase. Front Plant Sci 12:637685
5. Nakamura S, Abe F, Kawahigashi H et al (2011) A wheat homolog of MOTHER OF FT AND TFL1 acts in the regulation of germination. Plant Cell 23:3215–3229
6. Ito M, Hatta K, Terasawa Y et al (2021) Breeding of an extra-strong winter wheat variety with high pre-harvest sprouting tolerance, "Yumechikara 2020". Breed Res 23:37–43
7. Nishimura T, Mori M, Kamada T et al (2019) Detection of QTL for pre-harvest sprouting resistance and grain dormancy in highly sprouting-tolerant wheat. Breed Res 21:115–123. (In Japanese with English summary)
8. Chono M, Matsunaka H, Seki M et al (2013) Isolation of a wheat (*Triticum aestivum* L.) mutant in ABA 8′-hydroxylase gene: effect of reduced ABA catabolism on germination inhibition under field condition. Breed Sci 63:104–115
9. Fujita M (2019) Comparison of conditions and methods for testing ear germination. In: The 20th Anniversary Magazine of the Society for pre-harvest sprouting, pp 122–133. (in Japanese)
10. Nakamura S, Chono M, Stehno Z et al (2015) Diversification of the promoter sequences of wheat *Mother of FT and TFL1* on Chromosome 3A. Mol Breed 35:164
11. Nakatsu S (2000) High α-amylase activity in wheat (*Triticum aestivum* L.) grain from Hokkaido. Rep Hokkaido Res Org Agri Exp Stn 93: 6–18. (In Japanese with English summary)

INDEX

Naoto Kawakami and Kazuhiro Sato (eds.), *Seed Dormancy: Methods and Protocols*, Methods in Molecular Biology, vol. 2830, https://doi.org/10.1007/978-1-0716-3965-8,

Q

R

S

T

U

V

W

Y